Bioadhesion and Biomimetics

Bioadhesion and Biomimetics

From Nature to Applications

edited by
Havazelet Bianco-Peled
Maya Davidovich-Pinhas

Pan Stanford Publishing

Published by

Pan Stanford Publishing Pte. Ltd.
Penthouse Level, Suntec Tower 3
8 Temasek Boulevard
Singapore 038988

Email: editorial@panstanford.com
Web: www.panstanford.com

British Library Cataloguing-in-Publication Data
A catalogue record for this book is available from the British Library.

Bioadhesion and Biomimetics: From Nature to Applications

ISBN 978-981-4463-98-0 (Hardcover)
ISBN 978-981-4463-99-7 (eBook)

Printed in the USA

Contents

Preface xiii

SECTION I INTRODUCTION

1. Principles of Bioadhesion 3

John D. Smart

1.1 Introduction 4
1.2 The Principles of Adhesion 4
1.2.1 Theories of Adhesion 5
1.2.2 Forming the Adhesive Joint 7
1.2.2.1 Contact stage 7
1.2.2.2 The consolidation stage 8
1.3 Examples of Bioadhesion 8
1.3.1 Cell-to-Cell Adhesion 8
1.3.1.1 Cadherins 9
1.3.1.2 Immunoglobulins 9
1.3.1.3 Selectins 9
1.3.1.4 Integrins 10
1.3.2 Mucoadhesion 10
1.3.3 Other Medical Bioadhesives 14
1.3.3.1 Skin adhesives 14
1.3.3.2 Tissue sealants 14
1.3.3.3 Dental adhesives and cements 15
1.3.3.4 Bone cements and bone graft substitutes 16
1.4 Other Bioadhesives 16
1.5 Conclusion 17

2. Characterization of Bioadhesion **23**

Thakur Raghu Raj Singh, David S. Jones, Gavin P. Andrews, and Ravi Sheshala

2.1 Introduction 24
2.2 Methods of Bio-/Mucoadhesion Characterization 25
2.2.1 In vitro Testing 26
2.2.1.1 Methods measuring the force of detachment 26
2.2.1.2 Rheological measurement of mucoadhesion 31
2.2.1.3 Flow-through or falling liquid film method 34
2.2.1.4 Imaging techniques 35
2.2.1.5 Other in vitro tests 37
2.2.2 In vivo studies 41
2.3 Conclusion 42

SECTION II NATURAL ADHESIVES

3. Mussel Adhesives **49**

Hongbo Zeng, Qingye Lu, Bin Yan, Jun Huang, Lin Li, and Zhi Liao

3.1 Introduction 49
3.2 Mussel Adhesive System 50
3.2.1 Mussel Byssus 50
3.2.2 Biochemistry of Mussel Byssus 52
3.3 Understanding Interactions of Mussel Foot Proteins 54
3.3.1 Techniques for Studying Molecular Interactions of Mussel Foot Proteins 54
3.3.1.1 General surface analytical techniques 55
3.3.1.2 Atomic force microscopy 56
3.3.1.3 Surface forces apparatus 57
3.3.2 Protein–Substratum Interactions 60

3.3.2.1 Interaction between mussel foot proteins and mica 60
3.3.2.2 Interaction between mussel foot proteins and silica 62
3.3.2.3 Interaction between mussel foot proteins and TiO_2/metal 62
3.3.2.4 Interaction between mussel foot proteins and polymers 63
3.4 Protein–Protein Interactions in Byssal Plaque 64
3.4.1 The Interaction of Mussel Foot Protein 1 (mfp-1) in the Cuticle 64
3.4.1.1 Interactions between foot proteins in the plaque 66
3.5 Mussel-Inspired Materials 68
3.6 Adhesive/Sealants 69
3.7 Functional Coating 72
3.8 Conclusions 73

4. Gecko Adhesion **85**

Joseph C. Cremaldi, Kejia Jin, and Noshir S. Pesika

4.1 Introduction 86
4.2 The Gecko Pad 87
4.2.1 Hierarchical Tiers 87
4.2.2 Van der Waals Forces 88
4.2.3 Roughness Considerations 90
4.2.4 Force Generation and the Autumn Experiment 91
4.2.5 Force Results 91
4.2.6 Comparison with Whole Pad and Single Spatula Measurements 93
4.3 Attachment and Detachment 93
4.3.1 The Peel-Zone Model 95
4.3.2 Peel Modes 97
4.4 Self-Cleaning Mechanism 98
4.5 Non-Adhesive Unloaded State and Non-Self-Adherence 100

4.6 Substrates and the Environment 101
4.7 Conclusion 103

5. From Sand Tube to Test Tube: The Adhesive Secretion from Sabellariid Tubeworms 109

Elise Hennebert, Barbara Maldonado, Cécile Van De Weerdt, Mélanie Demeuldre, Katharina Richter, Klaus Rischka, Patrick Flammang

5.1 Introduction 110
5.2 Characterization of the Natural Adhesive System 111
5.2.1 Morphology of the Adhesive Cells 111
5.2.2 Morphology and Mechanics of the Cement 113
5.2.3 Cement Composition and Formation 115
5.3 Bio-Inspired Adhesive Polymers: Production, Characterization and Applications 118
5.3.1 Recombinant Proteins 119
5.3.2 Synthetic Peptides 120
5.3.3 Polymers 122

SECTION III BIOMIMETIC ADHESIVES

6. Adhesives and Coatings Inspired by Mussel Adhesive Proteins 131

Hao Meng, Yuan Liu, Morgan M. Cencer, and Bruce P. Lee

6.1 Introduction 131
6.2 Chemistry of DOPA Side Chain 133
6.2.1 Reversible Interactions 134
6.2.2 Covalent Bond Formation 135
6.3 Architectures of Synthetic Adhesive Polymers 136
6.3.1 Main Chain Functionalization 136
6.3.2 End Group Functionalization 137
6.4 Bioinspired Synthetic Adhesives 138
6.4.1 Injectable Tissue Adhesive 138
6.4.2 Nanocomposite Adhesive Hydrogel 140
6.4.3 Thin-Film Adhesive 142

6.4.4 Hybrid Gecko- and Mussel-Inspired Adhesive 143

6.5 Mussel-Inspired Antifouling Coatings 144

6.5.1 Graft-to Coating Method 145

6.5.2 Graft-From Coating Method 146

6.6 Polydopamine as a Versatile Multifunctional Anchoring Group 147

6.6.1 Polydopamine-Mediated Surface Modification 149

6.6.2 Metallization and Metal Ion Reduction 150

6.6.3 Polydopamine Capsule 150

6.7 Summary and Future Outlook 151

7. Algal Glue Mimetics 167

Ronit Bitton

7.1 Introduction 167

7.2 Algal Glue 168

7.2.1 Green Algae 168

7.2.2 Brown Algae 169

7.3 Biomimetic Algal Glue 172

8. Bio-Inspired Surfaces with Directional Adhesion 181

Luciano Afferrante

8.1 Introduction 181

8.2 Energy Release Rate 186

8.2.1 Detachment Occurring Perpendicularly to the Microwalls 186

8.2.2 Detachment Occurring Obliquely to the Microwalls 190

8.3 The Critical Moment for Detachment 190

8.3.1 Detachment Occurring Perpendicularly to the Microwalls 190

8.3.2 Detachment Occurring Obliquely to the Microwalls 194

8.4 Conclusions 195

SECTION IV MEDICAL APPLICATIONS

9. Bioadhesives as Surgical Sealants: A Review **203**

Leslie John Ray Foster

9.1 Introduction 203
9.2 Commercial Sealants 205
9.3 Fibrin-Based Sealants 206
9.3.1 Principles 206
9.3.2 Commercial Fibrin Sealants 207
9.3.3 Research and Development 208
9.4 Collagen-Based Sealants 209
9.4.1 Principles 209
9.4.2 Commercial Collagen Sealants 210
9.4.3 Research and Development 211
9.5 Gelatine-Based Sealants 211
9.5.1 Principles 211
9.5.2 Commercial Gelatine Sealants 212
9.5.3 Research and Development 213
9.6 Albumin-Based Sealants 214
9.6.1 Principles 214
9.6.2 Commercial Albumin Sealants 214
9.6.3 Research and Development 215
9.7 Chitosan-Based Sealants 216
9.7.1 Principles 216
9.7.2 Commercial Chitosan Sealants 217
9.7.3 Research and Development 217
9.8 Cellulose-Based Sealants 218
9.8.1 Principles 218
9.8.2 Commercial Cellulose Sealants 219
9.8.3 Research and Development 219
9.9 Summary and Future Trends 220

10. Bioadhesive Systems for Drug Delivery **235**

Ryan F. Donnelly and A. David Woolfson

10.1 Introduction 236

10.2 Routes of Administration for Bio- and Mucoadhesive Drug Delivery Systems 237
10.3 Mucoadhesive Polymers 239
10.4 Factors Affecting Mucoadhesion 242
10.4.1 Hydrophilicity 242
10.4.2 Molecular Weight 243
10.4.3 Cross-Linking and Swelling 243
10.4.4 Spatial Conformation 243
10.4.5 pH 244
10.4.6 Concentration of Active Polymer 244
10.4.7 Drug/Excipient Concentration 244
10.4.8 Other Factors Affecting Mucoadhesion 245
10.5 Delivery Systems 245
10.5.1 Oral Mucoadhesive Delivery Systems 245
10.5.2 Nasal Mucoadhesive Delivery Systems 248
10.5.3 Ocular Mucoadhesive Delivery Systems 250
10.5.4 Vaginal Mucoadhesive Delivery Systems 252
10.5.5 Rectal Mucoadhesive Delivery Systems 253
10.5.6 Cervical and Vulval Delivery Systems 254
10.5.7 Gastrointestinal Mucoadhesive Delivery Systems 255
10.6 Methods of Determining Mucoadhesion 256
10.6.1 Determination of Mucoadhesion in Polymer Gel Systems 256
10.6.2 Determination of Mucoadhesion in Film Systems 258
10.6.3 Tensile and Shear Tests 258
10.6.4 Peel Strength 259
10.7 Conclusion 260

11. Preventing Adhesion on Medical Devices **269**

Pilar Teixeira and Fernanda Gomes

11.2 Adhesion and Biofilm Formation on Medical Devices 271

11.3 Strategies to Prevent Adhesion on Medical Devices 273

11.3.1 Current Recommendations 273

11.3.2 Emergent Approaches 274

11.3.2.1 Surface topography modification 274

11.3.2.2 Surface coatings 275

11.3.2.3 Drug-delivery coatings 277

11.3.2.4 Antimicrobial peptides 278

11.3.2.5 Quorum-sensing molecules 279

11.3.2.6 Nanoparticles 280

11.3.2.7 Bacteriophages 281

11.3.2.8 Natural compounds 283

11.3.2.9 Electrical methods 284

11.4 Conclusions 285

Index 295

Preface

The term bioadhesion refers to the formation of interfacial bonds between two materials or surfaces, at least one of which is biological or biologically derived substance. An understanding of the fundamental mechanisms that govern bioadhesion is of great interest for researchers in various fields of science. One area of active research in this field focuses on natural adhesive materials produced by or extracted from plants, animals, fungi, and bacteria. Many of these bioadhesive materials act in Nature to permanently attach organisms such as plants, algae, and mussels to surfaces. Temporary bioadhesion events are also very common in Nature. These events typically involve adhesion organs allowing transient attachment of creatures such as frogs and geckos to surfaces. A second area of research in the field of bioadhesives focuses on biomimetic adhesives. These man-made materials aim to mimic the remarkable adherence capabilities of natural adhesives. To date, a variety of materials have been fabricated based on the lessons learned from Nature. Finally, studies in the field of biomedical engineering focus on bioadhesives applied as drug delivery vehicles or tissue adhesives. Bioadhesive drug delivery systems are mostly based on synthetic or natural polymers capable of bonding to mucosal surface. The term mucoadhesion may be used synonymously with bioadhesion to describe these systems. Tissue adhesives are useful in the surgical arena as a means to control bleeding and leaks of other body fluids.

Given the interdisciplinary nature of the discipline, this edited book has been designed to appeal to a broad spectrum of readers. It includes four sections addressing the latest developments relative to different aspects of Bioadhesion. The first section covers the principles and mechanisms governing the phenomena of biodahesion and describes experimental means to measure it. The second section focuses on natural adhesives, including "wet" adhesives produced and secreted by algae, mussels and send tube worms, and "dry" adhesion strategies utilized by geckos.

The third section deals with biomimetic adhesives and surfaces. Finally, the last section of this book is devoted to medical applications of adhesives.

This book gives a good background for any researcher dealing with the development of bioadhesives and we hope it would serve as a catalyst for further innovations in the field.

We would like to express our sincere thanks to our colleagues, the authors, who responded to our invitations and contributed to this edited book. Their collective expertise represents many years of industrial and academic experience in the field of Bioadhesion. We hope that this book would serve as a catalyst for further research and innovations as well as serve as a useful reference in the emerging field of bioadhesion.

Havazelet Bianco-Peled
Maya Davidovich-Pinhas

Section I

Introduction

Chapter 1

Principles of Bioadhesion

John D. Smart

School of Pharmacy and Biomolecular Sciences,
University of Brighton, Lewes Road, Brighton BN2 4GJ, UK

john.smart@brighton.ac.uk

Bioadhesion is said to occur when two materials, at least one of which is biological in nature, are held together for extended periods. Bioadhesion includes cell-to-cell adhesion, bacteria binding to surfaces, adhesion to mucous membranes and the use of adhesive materials in medical treatments (such as in drug delivery devices, tissue adhesives or wound dressings). There are two stages in the formation of the bioadhesive joint: The first is the contact stage, when two surfaces are brought together into intimate contact, and the second is the consolidation stage, when the adhesive interactions occur. There are several theories as to how such interactions arise, based around surface "wetting", macromolecular interpenetration, mechanical interlocking, adsorption, fracturing of the adhesive bond and electron transfer across an interface. The bioadhesion process is usually complex, and requires the application of more than one of these theories. Knowledge of the mechanism of bioadhesion will permit a greater understanding of the processes that happen in the natural world, and allow the development of more effective systems that can be used in medical therapies.

Bioadhesion and Biomimetics: From Nature to Applications
Edited by Havazelet Bianco-Peled and Maya Davidovich-Pinhas

ISBN 978-981-4463-98-0 (Hardcover), 978-981-4463-99-7 (eBook)
www.panstanford.com

1.1 Introduction

Bioadhesion may be defined as the state in which two materials, at least one of which is biological in nature, are held together for extended periods of time by interfacial forces. In medicine and pharmacy, the term "bioadhesion" refers to the adhesion of synthetic and biological macromolecules to a biological tissue. The biological substrate may be cells, bone, dentine or the mucus coating the surface of a tissue. If adhesive attachment is to a mucus coating, the phenomenon is sometimes referred to as mucoadhesion [1]. In many cases, in order to facilitate adhesion when two surfaces are brought together a third component (an adhesive or "glue") is required.

Many examples of bioadhesion exist in nature, and these include such diverse events as cell-to-cell adhesion within a living tissue, barnacles binding to rocks and bacteria binding to tooth enamel. It is apparent from mammalian anatomy and development that the associations of cells in epithelia, their attachment to basement membranes and the migrations of cells and projections of neurons all require selective adhesion of cells to one another and to extracellular matrices (ECMs), and special molecules are needed to facilitate this [2]. In healthcare, bioadhesives were first used as wound dressings, skin adhesives and denture fixatives [3]. Bioadhesives have become of interest in the development of medicines for their potential to optimize local drug delivery, by retaining a dosage form at the site of action (e.g. on the cervix or buccal mucosa), or systemic drug absorption by retaining the delivery system at the absorption site (e.g. the small intestine or nasal cavity). Bioadhesives could also be used as therapeutic agents in their own right, to coat and protect damaged tissues (lesions of the cornea or oral mucosa) or to act as lubricating agents (in the eye or oral cavity). Skin adhesives, tissue sealants, dental and bone adhesives/cements are also defined as bioadhesives.

1.2 The Principles of Adhesion

In conventional chemistry, it is considered that for adhesion to occur, molecules must bond across the interface. These bonds can arise in the following way [4].

Ionic bonds: Two oppositely charged ions attract each other via electrostatic interactions to form a strong bond (e.g. in a salt crystal).

Covalent bonds: Electrons are shared, usually in pairs, between the bonded atoms in order to "fill" the orbitals in both. These are also strong bonds.

Hydrogen bonds: A hydrogen atom, when covalently bonded to electronegative atoms such as oxygen or nitrogen, carries a slight positively charge and is therefore attracted to other electronegative atoms. The hydrogen can therefore be thought of as being shared, while the bond formed is generally weaker than ionic or covalent bonds.

Van-der-Waals bonds: Arise from dipole–dipole and dipole-induced dipole attractions in polar molecules, and dispersion forces with non-polar substances. These are some of the weaker forms of interaction.

Hydrophobic bonds (the hydrophobic effect): Indirect bonds (such groups only appear to be attracted to each other) that occur when non-polar groups are present in an aqueous solution. Water molecules adjacent to non-polar groups form hydrogen bonded structures, which lowers the system entropy. There is therefore an increase in the tendency of non-polar groups to associate with each other to minimize this effect. These are also considered to be some of the weakest interactions.

1.2.1 Theories of Adhesion

There are six general theories of adhesion, which have been adapted for the investigation of bioadhesion [5–7].

The Wetting theory: This is primarily applied to liquid systems and considers surface and interfacial energies. It involves the ability of a liquid to spread spontaneously onto a surface as a prerequisite for the development of adhesion. The spreading coefficient (S_{AB}) can be calculated from the surface energies of the solid and liquids using the equation:

$$S_{AB} = \gamma_B - \gamma_A - \gamma_{AB}, \tag{1.1}$$

where γ_A is the surface tension (energy) of the liquid A, γ_B is the surface energy of the solid B and γ_{AB} is the interfacial energy between the solid and liquid. S_{AB} should be positive for the liquid to spread spontaneously over the solid.

The work of adhesion (W_A) represents the energy required to separate the two phases, and is given by

$$W_A = \gamma_A + \gamma_B - \gamma_{AB} \quad (1.2)$$

The greater the individual surface energies of the solid and liquid relative to the interfacial energy, the greater the work of adhesion.

The electronic theory: This suggests that electron transfer occurs across contacting adhering surfaces due to differences in their electronic structure. This is proposed to result in the formation of an electrical double layer at the interface, with subsequent adhesion due to attractive forces.

The adsorption theory: This describes the attachment of adhesives on the basis of hydrogen bonding and van der Waals' forces. It has been proposed that these forces are the main contributors to the adhesive interaction. A subsection of this, the chemisorption theory, assumes an interaction across the interface occurs as a result of strong covalent bonding.

The diffusion theory: This theory describes the interdiffusion of polymers chains across an adhesive interface. This process is driven by concentration gradients and is affected by the available molecular chain lengths, the compatibility of the two polymers and their mobilities. The depth of interpenetration depends on the diffusion coefficient and the time of contact. Sufficient depth of penetration creates a semi-permanent adhesive bond.

The mechanical theory: This assumes that adhesion arises from an interlocking of a liquid adhesive (on setting) into irregularities on a rough surface. However, rough surfaces also provide an increased surface area available for interaction along with an enhanced viscoelastic and plastic dissipation of energy during joint failure, which are thought to be more important in the adhesion process than a mechanical effect [7].

The fracture theory: This differs a little from the other five in that it relates the adhesive strength to the forces required for the detachment of the two involved surfaces after adhesion. This assumes that the failure of the adhesive bond occurs at the interface. However, failure normally occurs at the weakest component, which is typically a cohesive failure within one of the adhering surfaces.

The bioadhesive process can be complex and often involves more than one of the above.

1.2.2 Forming the Adhesive Joint

There are effectively two stages in the formation of an adhesive joint [8] (Fig. 1.1).

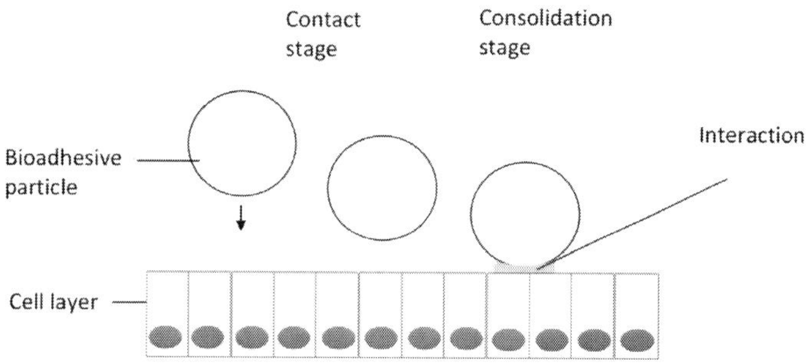

Figure 1.1 The two stages in the establishment of a bioadhesive bond.

1.2.2.1 Contact stage

In the contact stage the surfaces have initially to come together to form an intimate contact. This is facilitated in many cases by two surfaces being physically brought together, e.g. placing and holding an adhesive patch onto the skin surface. In other occasions, forming the initial contact may be a critical step in the adhesive process, such as in the adhesion of bacteria to surfaces or the deposition of a particle within in the nasal cavity or bronchi [9].

The principles of the DLVO theory, described in the 1940s by Derjaguin and Landau and separately by Verwey and Overbeek to explain the stability of colloids [9], have been used to describe the physicochemical processes involved in the adsorption of small particles (such as bacteria) onto surfaces [10,11]. In suspension, a particle will be constantly moving due to Brownian motion and further movement will occur in vivo due to the flow of liquids within a body cavity and body movements such as peristalsis. When a particle approaches a surface, it will experience both attractive and repulsive forces. Attractive forces arise from van-der-Waals' interactions, surface energy effects and electrostatic interactions if the surface and particles carry opposite charges. Repulsive forces arise from osmotic pressure effects as a result of the interpenetration of the electrical double layers, steric effects and also electrostatic interactions when the surface and particle carry the same charge. The relative strength of these opposing forces will depend on the nature of the particle, the aqueous environment, and the distance between the particle and surface. For example, the smaller the particle, the greater the surface-area-to-volume ratio and therefore the greater the attractive forces. Particles can be weakly held at a secondary minimum (circa 10 nm separation), a region where the attractive forces are balanced by the repulsive forces allowing the particles to be easily dislodged. For stronger adsorption to occur, particles have to overcome a repulsive barrier (the potential energy barrier) to get closer to the surface (circa 1 nm). If this barrier is sufficiently small, or if the particle has sufficient energy, then adsorption into the primary minimum can occur. This type of adsorption would be required to allow a strong adhesive bond to form. This situation is complicated in vivo as the surface in question is a soft biological surface rather than a flat and rigid solid, and the particles may become hydrated and/or coated with biomolecules, significantly altering their physicochemical properties [12–14].

The adhesive interaction necessary to retain a dosage form may only need to be weak if the forces promoting displacement are also small, such as for a small particle in the unstirred water layer at the surface of the gastrointestinal mucosae [15,16], or become lodged in these surface folds and crevasses of the gastrointestinal tract. This might explain how apparently inert materials have been reported to be "bioadhesive" [17–19]. However, in order for the formation of a stable adhesive joint, a second consolidation stage is required.

1.2.2.2 The consolidation stage

In this stage adhesive interactions need to occur to establish a stable mucoadhesive joint capable of withstanding those dislodgement forces likely to be encountered in a biological environment, using the mechanisms outlined above. Typically these adhesive interactions are complex and involve more than one mechanism.

1.3 Examples of Bioadhesion

1.3.1 Cell-to-Cell Adhesion

The specificity of cell adhesion arises from the expression and interactions among a range of adhesion receptors on cell surfaces, and induction relies on ligands binding to receptors, cell–cell contacts and cell–matrix adhesion [2]. These arise typically be secondary chemical bonding. The following groups of molecules have been identified that facilitate this adhesion.

1.3.1.1 Cadherins

Cadherins, a class of type-1 transmembrane proteins, mediate calcium ion dependent (hence the name) homophilic (like-with-like) adhesion between cells through the most distal cadherin repeats [20]. There are several subgroups of cadherins.

The "classic" cadherins share homologous cytoplasmic domains that link to the actin cytoskeleton. Both structural and functional analyses suggest that the functional unit is a dimer. As with other adhesion receptors, clustering of cadherins is important for their functions, and multiple dimer to dimer interactions are believed to provide sufficient local avidity to mediate cell-to-cell adhesion.

Desmosomal cadherins (desmocollins, desmogleins), although related to classic cadherins in their extracellular domains, they have distinct cytoplasmic domains that link to intermediate filaments.

Other subclasses of the cadherin superfamily are known as protocadherins, and these typically have six cadherin repeats.

1.3.1.2 Immunoglobulins

An immunoglobulin (or antibody), is a large Y-shaped protein that is typically used by theimmune system to identify and neutralize foreign objects such as bacteria and viruses. The antibody recognizes a unique part of the foreign target, called an antigen. Immunoglobulins are characterized by the presence of varying numbers of Immunoglobulin-related domains [21]. Like cadherin domains, these are sandwiches of two sheets held together by hydrophobic interactions. The immunoglobulin superfamily is diverse, numbering well over 100 in vertebrates. Numerous molecules with one or more immunoglobulin domains can play roles in cell-to-cell interactions in the immune system and elsewhere.

1.3.1.3 Selectins

Another well-studied group of cell adhesion receptors comprises the selectins and their counter-receptors [22]. Binding is through the C-type lectin domain in the selectin, which recognizes specific carbohydrate groupings in the counter receptor/ligand. Unlike cadherins and Immunoglobulin-SF members, selectins are currently known only in cells of the vertebrate circulation (endothelium and blood cells), although other lectins are widely distributed. Selectins and their ligands play an important role in the adhesion of leukocytes to endothelium, where their cooperation with integrins and Immunoglobulin-SF receptors is one of the best-understood examples of cell adhesion specificity, which arises from tightly regulated display and interaction among a limited number of receptors.

1.3.1.4 Integrins

Integrins are transmembrane receptors that mediate the attachment between a cell and its surroundings, such as other cells or the extracellular matrix. Most integrins are predominantly or exclusively receptors for extra cellular matrix proteins such as fibronectins, laminins and collagens, but a few also play important roles in cell adhesion, most notably of leukocytes, where they bind to counter-receptors of the immunoglobulin superfamily or a cadherin.

1.3.2 Mucoadhesion

This typically describes a situation where a material adheres to a mucous membrane, more specifically interacting with the mucus at the surface. Mucous is a lubricant, designed to stop materials adhering, so a mucoadhesive will need to change the nature of this outer layer if it is to promote adhesion. In much of the literature the interest in bio/mucoadhesion is for their potential to enhance drug therapy, by locating and retaining a drug delivery system on a mucosal surface, such as particles in the gastrointestinal tract, in order to either optimize local therapy or enhance absorption.

The most widely investigated group of mucoadhesives are hydrophilic macromolecules containing numerous hydrogen bond forming groups [23–27]. The presence of hydrogen bond forming groups (hydroxyl, carboxyl or amine) on the molecules favours adhesion. They are sometimes referred to as "wet adhesives" in that they require moisture to become activated and will adhere non-specifically to many surfaces [28]. In an aqueous environment (such as within the human body) they may overhydrate to form a slippery mucilage, and this can be responsible for adhesive joint failure. Like typical hydrocolloid glues, if the formed adhesive joint is allowed to dry then they can form very strong adhesive bonds. Interestingly these properties are similar to those of mucus itself. Typical examples of this type of mucoadhesives are carbomers, chitosan and the cellulose derivatives (Fig. 1.2).

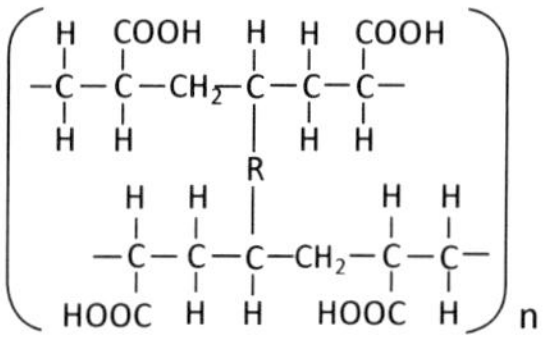

(a) Poly(acrylic acid), R = allyl sucrose or allyl pentaerythritol (Carbopols); or divinyl glycol (polycarbophil)

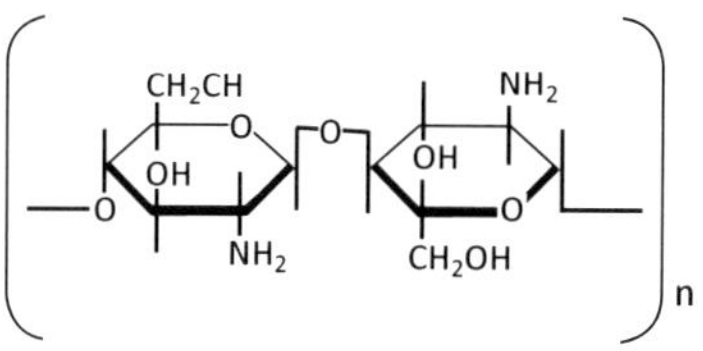

(b) Chitosan

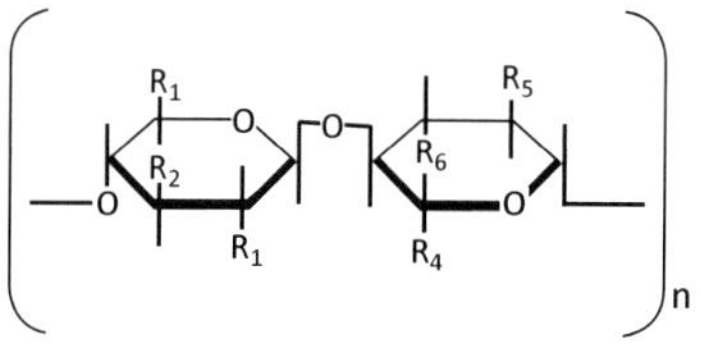

(b) Cellulose derivatives e.g.
Sodium carboxylmethylcellulose
R1, R4 = CH_2OH; R2, R3, R5 = OH;
R6 = $OCH_2CO_2^-Na^+$
Hydroxypropylmethylcellulose
R1 = CH_2OCH_3; R_2 = OH, R3 = $OCH_2CHOHCH_3$;
R4 = CH_2OH; R5, R6 = OCH_3

Figure 1.2 The structure of some common mucoadhesive polymers.

In considering how mucoadhesion arises, a whole range of "scenarios" are possible depending in particular on whether the formulation is a solid (tablet or patch), semisolid (an ointment or gel) or liquid (eye drops or mouth washes. In the latter two, we tend to talk about retention rather than adhesion. If the liquid wets the surface then retention depends largely on the rheology of the liquid/semisolid and the environmental challenges it is exposed to. Components of the formulation can also adsorb onto the surface, in a similar fashion to mucin glycoprotein. Therefore, in this section, mucoadhesion arising when dry (initially) solid materials are brought into contact with a mucosal surface.

Once placed into contact with a mucosal surface, for successful mucoadhesion to occur, strong or prolonged adhesion is usually required, for example with larger formulations exposed to stresses such as blinking or mouth movements. Once activated by the presence of moisture, mucoadhesive materials adhere most strongly to solid dry surfaces [29]. Moisture will effectively plasticize the system allowing mucoadhesive molecules to become free, conform to the shape of the surface, and bond predominantly by weaker van-der-Waal and hydrogen bonding, although ionic interactions can also occur in some cases. The mucoadhesive bond is by nature very heterogeneous making it extremely difficult to use spectroscopic techniques to identify the type of bonds and groups involved although hydrogen bonds have been identified as being important [30–32]. When undertaking tensiometer studies of mucoadhesion the high affinity of materials like carbomers for water almost appears to have a "suction-like" effect, which holds the formulation onto a solid surface [29]. For surfaces with only a thin mucus layer, a dry mucoadhesive polymer, by extracting the water component of the gel, will almost certainly dehydrate and collapse this [33]. However, when a substantial mucus layer is present, its lubricant/anti-adherent properties will need to be overcome to allow strong adhesion. Here the adhesive joint can be considered to contain three regions, the mucoadhesive, the mucosa and an interfacial region, which consists at least initially of mucus. Adhesive joint failure occurs at the weakest region of the adhesive joint, and in this case this would be expected to be the mucus. To achieve strong adhesion, a change in the physical properties of the mucus layer is therefore required (gel "strengthening").

There are essentially two theories as to how this gel strengthening/consolidation occurs. One is based on a macromolecular interpenetration effect, which has been dealt with on a theoretical basis [7]. In this theory, analogous to the diffusion theory described by Voyutskii [33] for compatible polymeric systems, the mucoadhesive molecules interpenetrate and bond by secondary interactions with mucus glycoproteins (Fig. 1.3). The second theory is the dehydration theory [32]. When a material capable of rapid gelation in an aqueous environment is brought into contact with a mucus gel, water moves rapidly between gels until equilibrium is achieved.

A polyelectrolyte gel, such as a poly(acrylic acid) will have a strong affinity for water, therefore a high "osmotic pressure" and a large swelling force [34,35]. When brought into contact with a mucus gel it will rapidly dehydrate that gel and force intermixing and consolidation of the mucus joint until equilibrium is reached (Fig. 1.4). The rapid formation of the adhesive bond suggests that the dehydration theory is the more likely explanation of mucoadhesion.

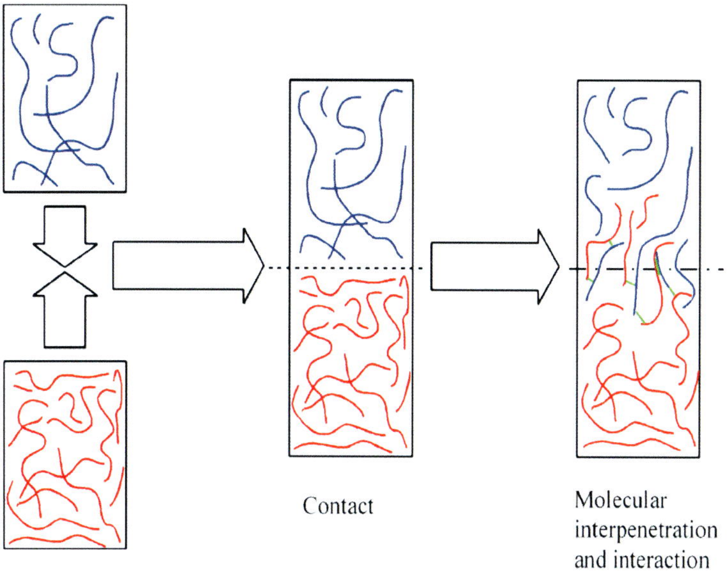

Figure 1.3 The interpenetration theory of mucoadhesion; the stages in the interaction between a mucoadhesive polymer and mucin glycoprotein in a mucus gel.

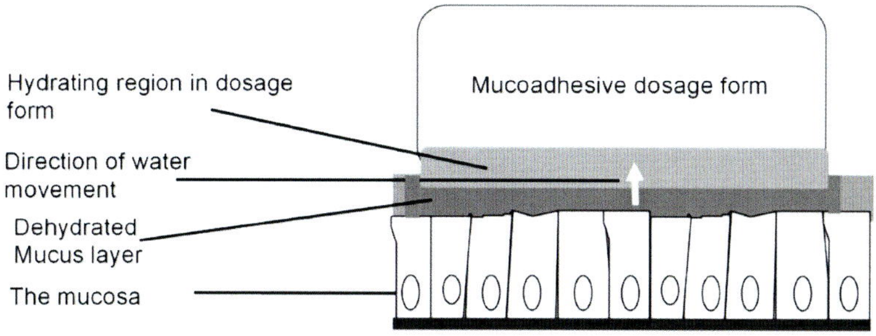

Figure 1.4 The dehydration theory of mucoadhesion.

1.3.3 Other Medical Bioadhesives

1.3.3.1 Skin adhesives

These materials differ greatly from mucoadhesives in that they do not need water to become activated (in fact, water may inhibit the adhesion process) and must therefore possess inherent "stickiness" (tack) [36]. They must be sufficiently plastic and hydrophobic to interact with and adhere to the skin surface on application. Rubber (polyisoprene) was the first commercial skin adhesive. Today the three main classes of adhesives are based on polyisobutylenes (or other synthetic rubbers), acrylate polymers (such as 2-ethylhexyl-, butyl- and/or octyl-acrylate copolymers) or silicone polymers (such as polydimethyl and polydimethyldiphenyl silicone).

Hydrocolloid pressure sensitive adhesives contain a combination of the water-soluble polymers ("wet adhesives") similar to those used for mucoadhesive drug delivery and the pressure sensitive adhesives described above [37]. They are used in wound dressings as they have both adhesive and absorbent properties. A typical early formulation would contain a continuous phase of a low-molecular-weight polyisobutylene (circa 40%) into which is dispersed combinations of hydrocolloids such as pectin, sodium carboxymethyl cellulose and gelatine. Later versions use a continuous phase of thermoplastic copolymers such as styrene-isoprene-styrene into which the hydrocolloid is dispersed, and could also contain a range of other materials such as tackifiers and antioxidants, cross-linking agents and mineral oils (e.g. Duoderm® or Sigma Dress®).

1.3.3.2 Tissue sealants

Adhesives for tissue repair are used in surgery as an alternative to staples, sutures or surgical tapes [38] as they are relatively easily applied, acceptable, biodegradable, relatively inexpensive, seal more surely, reduce infection risk and virtually eliminate scarring, although in some cases may be less mechanically strong. There are several types of tissue sealants.

Fibrin sealants (e.g. Tisseel®) are composed primarily of the proteins fibrinogen and thrombin. These are the building blocks of a blood clot and on mixing work by mimicking the final stage of the body's natural clotting mechanism. The formed fibrin clot is both adhesive and haemostatic, so are particularly applicable to sealing bleeding wounds. Some variants are only activated by an external source, such as light, thus allowing the sealant to be activated only at the site of application.

Cyanoacrylates, used as tissue adhesives in surgical procedures essentially consist of cyanoacrylate monomer which on contact with trace amounts of water or bases can polymerize and binds surfaces together (Fig. 1.5). The original butyl cyanoacrylates "superglues" have now been replaced by octyl cyanoacrylates (Dermabond®) that are mechanically stronger, more stable, more flexible with less toxicity.

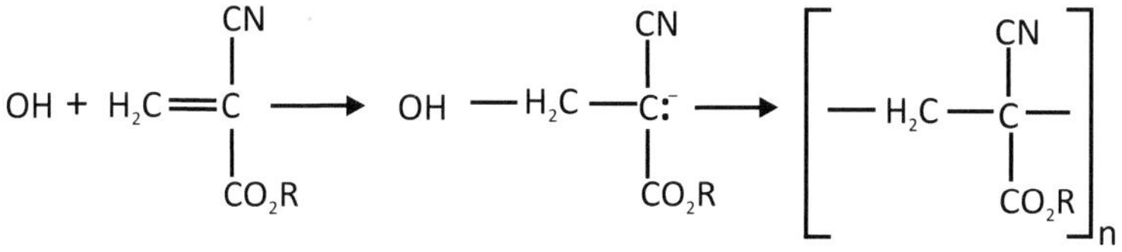

Figure 1.5 Cyanoacrylate superglues, their polymerization reaction (where R = butyl or octyl groups).

The protein albumin, when cross-linked with gluteraldehyde, forms a strong adhesive for internal surgery, capable of withstanding high pressures (BioGlue®). The protein collagen can also be used as the matrix to which other materials such as fibrin, thrombin or polyethylene glycols can attach. Derivatives of polyethylene oxide can be used as swellable hydrogels that can rapidly arrest bleeding (Co-seal®, Co-stasis®).

1.3.3.3 Dental adhesives and cements

Dental adhesives and cements adhere to the dental enamel and dentine of teeth in order to secure orthodontic devices. A range of materials are available, the choice depending on the required properties for the procedure to be undertaken [39]. Examples are zinc phosphate (e.g. Tenacin®) cement, which is based on the neutralization reaction of zinc oxide with phosphoric acid in an aqueous environment to form solid zinc phosphate; zinc polycarboxylate (e.g. Durelon®) cement, the reaction product of zinc oxide with aqueous polyacrylic acid, which forms a cement that adheres to the calcium present in dentine and enamel; glass ionomer cements (e.g. Fuji®), consisting of polyacrylic or polymaleic acid mixed with an ion-leachable aluminosilicate glass, that react together in water to form a cement that will also adhere to enamel or dentine and resin adhesives (e.g. Panavia®), consisting of resin monomers (such as the acrylate phosphate esters) that can be activated by light, initiators or a combination of both, and also often contain inert fillers. They cure purely through a polymerization process so contain little water and are less brittle and more fracture resistant than cements.

1.3.3.4 Bone cements and bone graft substitutes

Bone filling cements have been used in surgery for the fixation of prostheses such as in total hip replacements for over 40 years [40, 41]. Examples are:

The polymethyl methacrylates (PMMA) (e.g. Simplex P®). They are prepared by the polymerization of methylmethacrylate monomer, and a typical cement would contain monomer, an initiator for polymerization (e.g. benzoyl peroxide), a radio-opaque substance (e.g. barium sulphate) and oligomers of PMMA.

Calcium phosphate (e.g. Norion SRS®), in the form hydroxyapatite is a key component of human bone. These cements are easily accepted by the body and are sometimes used instead of bone-graft surgery. They consists of a powder to which water is added to form a paste which hardens to form a material structurally similar to bone, which can then be integrated into the new growth.

Bioactive glasses (NovaBone®) consist of sodium oxide, calcium oxide, phosphorous pentoxide, and silicon dioxide in varying proportions, that are solid and non-porous. They form a strong bond with bone as a result of a gel layer containing calcium and phosphate ions being generated at the surface on exposure to physiological aqueous solutions forming hydroxyapatite.

Glass ionomers (e.g. Fugi IX®), calcium sulphate (plaster of Paris), aluminium oxide (e.g. alumina ceramic) and synthetic hydroxyapatite (Calcitite®) have also been used as synthetic bone grafts.

1.4 Other Bioadhesives

Lectins are defined as proteins or glycoproteins capable of specific recognition of and reversible binding to carbohydrate moieties of complex glycoconjugates, without altering the covalent structure of any of the recognized glycosyl ligands [42]. The group of cell-to-cell adhesive molecules, the selectins, fall within the lectin group.

Although lectins are structurally diverse, it is possible to group many of them into distinct families of homologous proteins that share common structural properties [43]. The largest and best characterized belong to the family of legume lectins, which show extensive homologies in their primary sequences and express three-dimensional structural similarities. The exact physiological role of lectins is not fully understood but they are implicated in many cell recognition and adhesion processes. For example "lectin like" molecules that bind to cell surface glycoconjugates have been shown to be important in the adhesion of microorganisms to mucosal surfaces [44]). This therefore raises the possibility of using lectins to "anchor" a particle or macromolecular carrier to a specific target tissue. Many epithelial surfaces contain endogenous lectins that provide the opportunity to target with a system containing the appropriate hapten (binding) sugar [45], and this approach has been investigated to target polymers within the gastrointestinal tract [46].

Muscle glue protein has also been investigated as a bioadhesive material.

1.5 Conclusion

Bioadhesion is an important phenomenon and has been used by nature to mediate a range of physiological processes, and is of interest in medicine for a variety of therapeutic purposes. The principles behind bioadhesion are based on those encountered in the adhesive literature generally, the mechanism depending on the nature and environment of the two adhering surfaces.

References

1. Gu, J. M., Robinson, J. R., and Leung, S. H. S. (1988). Binding of acrylic polymers to mucin/epithelial surfaces: Structure property relationships, *Crit. Rev. Ther. Drug Carrier Syst.*, **5**, 21–67.
2. Hynes, R. O. (1999). Cell adhesion: Old and new questions, *Trends in Biochem. Sci.*, **24**, M33–M37.

3. Smart J. D. (2004). *Bioadhesion*, eds. Wnek G. E., and Bowlin, G. L. Encyclopaedia of Biomaterials and Biomedical Engineering (Marcel Dekker, New York), pp. 62–71.
4. Laidler, K. J., Meiser, J. H., and Sanctuary, B. C. (2003). *Physical Chemistry 4th Ed.* (Houghton Mifflin Company, Boston USA).
5. Ahuja, A., Khar, R. K., and Ali, J. (1997). Mucoadhesive drug delivery systems, *Drug Dev. Ind. Pharm.*, **23**, 489–515.
6. Mathiowitz, E., and Chickering D. E. (1999). *Definitions, Mechanisms and Theories of Bioadhesion*, eds. Mathiowitz, E., Chickering, D. E., and Lehr, C.-M. Bioadhesive Drug Delivery Systems: Fundamentals, Novel Approaches and Development, (Marcel Decker, New York), pp. 1–10.
7. Peppas, N. A., and Sahlin, J. J. (1996). Hydrogels as mucoadhesive and bioadhesive materials: A review, *Biomaterials*, **17**, 1553–1561.
8. Wu, S. (1982). *Polymer Interface and Adhesion*. (Marcel Dekker Inc. New York).
9. Florence, A. T., and Attwood, D. (1997). *Physicochemical Principles of Pharmacy*, 3rd ed (Palgrave Ltd, Basingstoke, UK).
10. Tunney, M. M., Gorman, S. P., and Patrick, S. (1996). Infection associated with medical devices, *Revs. Med. Microbiol.*, **7**, 195–205.
11. Poortinga, A. T., Bos, R., Norde, W., and Busscher H. J. (2002). Electrical double layer interactions in bacterial adhesion to surfaces. *Surf. Sci. Rep.*, **47**, 1–32.
12. He, P., Davis, S. S., and Illum, E. (1998). In-vitro evaluation of mucoadhesive properties of chitosan microspheres, *Int. J. Pharm.*, **166**, 75–88.
13. Degim, Z., and Kellaway, I. W. (1998). An investigation of the interfacial attraction between poly(acrylic acid) and glycoprotein, *Int. J. Pharm.*, **175**, 6–16.
14. Tur, K. M., and Ch'ng, H.-S. (1998). Evaluation of possible mechanisms of mucoadhesion, *Int. J. Pharm.*, **160**, 61–74.
15. Ashford, M. (2007). *The Gastrointestinal Tract: Physiology and Drug Absorption*, ed. Aulton, M. E., Pharmaceutics, the design and manufacture of medicines (Churchill Livingstone, Elsevier, UK), pp. 270–285.
16. Marriott, C. (2007). *Rheology*, ed. Aulton, M. E., Pharmaceutics, the design and manufacture of medicines (Churchill Livingstone, Elsevier, UK), pp. 42–58.

17. Thairs, S., Ruck, S., Jackson, S. J., Steele, R. J. C., Feely, L. C., Washington, C., and Washington, N. (1998). Effect of dose size, food and surface coating on the gastric residence and distribution of ion exchange resin. *Int. J. Pharm.*, **176**, 47–53.

18. Takeuchi, H., Yamamoto, H., and Kawashima, Y. (2001). Mucoadhesive microparticulate systems for peptide drug delivery, *Adv. Drug Deliv. Rev.*, **47**, 39–54.

19. Jackson, S. J., Bush, D., Washington, N., and Perkins, A. C., (2000). Effect of resin surface charge on gastric mucoadhesion and residence time of cholestyramine, *Int. J. Pharm.*, **205,** 173–181.

20. Junghans, D., Haas, I. G., and Kemler, R. (2005). Mammalian cadherins and protocadherins: About cell death, synapses and processing, *Curr. Opin. Cell Biol.*, **17**, 446–452.

21. Walsh, F. S., and Doherty, P. (1997). Neural cell adhesion molecules of the immunoglobulin superfamily: Role in axon growth and guidance. *Annu. Rev. Cell Dev. Biol.*, **13**, 425–456.

22. McEver, R. P. (2002). Selectins: Lectins that initiate cell adhesion under flow. *Curr. Opin. Cell Biol.*, **14**, 581–586.

23. Smart, J. D., Kellaway, I. W., and Worthington H. E. C. (1984). An in-vitro investigation of mucosa-adhesive materials for use in controlled drug delivery. *J. Pharm. Pharmacol.*, **36**, 295–299.

24. Chen, J. L., and Cyr, G. N. (1970). *Compositions Producing Adhesion Through Hydration*, ed. Manly R. S., Adhesion in Biological Systems (Academic Press, New York), pp. 163–181.

25. Harding, S. E., Davis, S. S., Deacon, M. P., and Fiebrig, I.(1999). Biopolymer mucoadhesives. *Biotechnol. Genet. Eng. Revs.*, **16**, 41–85.

26. Lee, J. W., Park, J. H, and Robinson J. R. (2000). Bioadhesive-based dosage forms: The next generation, *J. Pharm. Sci.*, **89**, 850–866.

27. Smart, J. D. (1993). Drug delivery using buccal adhesive systems, *Adv. Drug Deliv. Revs*, **11**, 253–270.

28. Ben Zion, O., and Nussinovitch, A. (1997). Physical properties of hydrocolloid wet glues, *Food Hydrocol*, **11**, 429–442.

29. Mortazavi, S. A., and Smart, J. D. (1995). An investigation of some factors influencing the in-vitro assessment of mucoadhesion, *Int. J. Pharm.*, **116**, 223–230.

30. Sriamornsak, P., Wattanakorn, N., Nunthanid J., and Puttipipatkhachorn, S. (2008). Mucoadhesion of pectin as evidence by wettability and chain interpenetration, *Carbohyd. Polymers*, **74**, 458–467.

31. Patel, M. M., Smart, J. D., Nevel, T. G., Ewen, R. J., Eaton, P. J., Tsibouklis, J. (2003). Mucin/polyacrylic acid interations: A spectroscopic investigation of mucoadhesion, *Biomacromolecules*, **4**, 1184–1190.

32. Smart, J. D. (1999). *The Role of Water Movement and Polymer Hydration in Mucoadhesion*, eds. Mathiowitz, E., Chickering, D. E., and Lehr, C.-M., Bioadhesive Drug Delivery Systems: Fundamentals, Novel Approaches and Development (Marcel Decker, New York), pp. 11–23.

33. Voyutskii, S. S. (1963). *Autoadhesion and Adhesion of High Polymers*, John Wiley and Sons/Interscience, New York.

34. Silberberg-Bouhnik, M., Ramon, O., Ladyzhinski, I., Mizrahi, S., and Cohen, Y. (1995). Osmotic deswelling of weakly charged poly(acrylic acid) solutions and gels, *J. Polymer Sci.*, **33**, 2269–2279.

35. Khare, A. R., Peppas, N. A., Massimo, G., Columbo, P. (1992). Measurement of the swelling force in ionic polymer networks, I. Effect of pH and ionic content, *J. Contr. Rel.*, **22,** 239–244.

36. Horstmann, M., Muller, W., and Asmussen, B. (1999). *Principles of Skin Adhesion and Methods for Measuring Adhesion of Transdermal Systems*, eds. Mathiowitz, E., Chickering, D. E., and Lehr, C.-M. Bioadhesive Drug Delivery Systems: Fundamentals, Novel Approaches and Development (Marcel Decker, New York), pp. 175–196.

37. Lipman R. (1999). Hydrocolloid PSAs: New formulation strategies. *Medical Device & Diagnostic Industry Magazine*, 32–150.

38. Sharma, A., Kaur, R., Kumar, S., Gupta, P., Pandov, S., Patnaik, B., and Gupta, A. (2003). Fibrin glue versus N-butyl-2-cyanoacrylate in corneal perforations, *Ophthalmol.*, **110**, 291–298.

39. Ewoldsen, N., and Demke, R. S. (2001). A review of orthodontic cements and adhesives, *Amer. J. Orthodont. Dentofac. Orthoped.*, **120**, 45–48.

40. DiMaio, F. R. (2002). The science of bone cement, a historical review. *Orthoped.*, **25**, 1399–1407.

41. Moore, W. R., Graves, S. E., and Bain, G. I. (2001). Synthetic bone graft substitutes, *Aust. New Zeal. J. Surg.*, **71**, 354–361.

42. Lehr, C.-M. (2000). Lectin-mediated drug delivery: The second generation of bioadhesives, *J. Contr. Rel.*, **65**, 19–29.

43. Woodley, J. (2001). Bioadhesion. New Possibilities for drug administration, *Clin. Pharmacokinet.*, **40**, 77–84.

44. Easson, J. H., Haltner, E., and Lehr, C.-M. (1999). Bacterial invasion factors and lectins as second generation bioadhesives, eds. Mathiowitz, E., Chickering, D. E., and Lehr, C.-M. Bioadhesive Drug

Delivery Systems: Fundamentals, Novel Approaches and Development (Marcel Decker, New York) 409–432.

45. Ponchel, G., and Irache, J.-M. (1998). Specific and non-specific bioadhesive particulate systems for oral delivery to the gastrointestinal tract, *Adv. Drug Deliv. Revs.*, **34**, 191–219.

46. Rihova, B., Rathi, R. C., Kopeckova, P., and Kopecek, J. (1992). In-vitro bioadhesion of carbohydrate containing *N*-(2-hydroxypropyl)-methacrylamide copolymers to the GI tract of guinea pigs, *Int. J. Pharm.*, **87**, 105–116.

Chapter 2

Characterization of Bioadhesion

Thakur Raghu Raj Singh,[a] David S. Jones,[a] Gavin P. Andrews,[a] and Ravi Sheshala[b]

[a]*School of Pharmacy, Queen's University Belfast,*
Medical Biology Centre, 97 Lisburn Road, Belfast BT9 7BL, UK
[b]*Department of Pharmaceutical Technology, School of Pharmacy,*
International Medical University, Bukit Jalil, 57000, Kuala Lumpur, Malaysia

r.thakur@qub.ac.uk

Bioadhesion is defined as the binding of a natural or synthetic polymer material to a biological substrate. Over the past few decades, bioadhesive-based drug delivery systems have received a great deal of attention. Bioadhesive dosage forms are primarily designed to attach and enable prolonged retention at the site of application, thus providing a controlled rate of drug release for improved therapeutic outcome. Most commonly these dosage forms are applied to the mucosal tissues of the body such as buccal, nasal, ocular, gastro, vaginal, and rectal mucosa. This mode of localized and/or systemic drug delivery will provide improved therapeutic benefits unlike other routes of drug delivery. For example, application of dosage forms to mucosal surfaces may be of benefit to drug molecules that are not amenable to the oral route, such as those that undergo acid degradation or extensive first-

Bioadhesion and Biomimetics: From Nature to Applications
Edited by Havazelet Bianco-Peled and Maya Davidovich-Pinhas

ISBN 978-981-4463-98-0 (Hardcover), 978-981-4463-99-7 (eBook)
www.panstanford.com

pass metabolism. The mucoadhesive ability of a dosage form is dependent upon a variety of factors, including the nature of the mucosal tissue and the physicochemical properties of the polymeric formulation. The bioadhesive-based drug delivery systems can be formulated into different pharmaceutical dosage forms namely as gels, ointments, tablets, patches, films, and microspheres. It is, therefore, essential that these formulations bond/attach to the mucosal surface for prolonged time to provide efficient drug delivery. Formulation scientist have designed and tested a range of techniques and set-ups that are now commonly used to select candidate bioadhesive drug delivery platforms. This Chapter provides an overview of the various in vitro, ex vivo and in vivo techniques that have been extensively used in the characterization of bio-/mucoadhesion drug delivery systems.

2.1 Introduction

The primary aim of a bioadhesive-based drug delivery system is to prolong the residence time at the site of application. Prolonged residence time is advantageous since it will provide effective local drug delivery at the site of application for long-term. Bioadhesive-based drug delivery systems are available in a wide range of pharmaceutical dosage forms such as gels, creams, tablets, patches, and films. Depending upon the type of dosage form different routes such as oral, rectal, vaginal, gastrointestinal tract, ocular or nasal are used to deliver bioadhesive-based delivery system. Consequently, it is essential to characterize polymers bioadhesive property to a given biological tissue samples or its substitute. Characterizing provides vital information during pre-formulation development process, which will enable formulation scientists to select a suitable bioadhesive dosage form for drug delivery applications. Therefore, in this Chapter, we have discussed various set-ups that are commonly used in in vitro and in vivo characterizing of bioadhesive-based polymeric drug delivery systems. Importantly, we have discussed a vast number of in vitro techniques that are most commonly used in the pre-formulation characterization of a bio-/mucoadhesive drug delivery system.

2.2 Methods of Bio-/Mucoadhesion Characterization

Adhesion can be defined as a process of "fixing" of two surfaces to one another [1]. When adhesion occurs in a biological setting it is often termed "bioadhesion". Bioadhesion can be defined as the binding of a natural or synthetic polymer to a biological substrate. When this substrate is a mucous layer, the term "mucoadhesion" is often used [2]. Mucoadhesion has been widely promoted as a way of achieving site-specific drug delivery through the incorporation of mucoadhesive hydrophilic polymers within pharmaceutical formulations (e.g. films) along with the active pharmaceutical ingredient (API). Mucoadhesive-based formulations allow localized delivery of the API that will be released close to the site of action with a consequent enhancement of bioavailability [3]. Bio-/mucoadhesive drug delivery systems provide several advantages, for example:

As a result of adhesion and intimate contact, the formulation stays longer at the delivery site improving API bioavailability using lower API concentrations for disease treatment.

The use of specific bioadhesive molecules allows for possible targeting of particular sites or tissues, for example the gastrointestinal (GI) tract.

Increased residence time combined with controlled API release may lead to lower administration frequency.

The avoidance of first-pass metabolism.

Additionally significant cost reductions may be achieved and dose-related side effects may be reduced due to API localization at the disease site [3].

Advantages of mucoadhesive-based drug delivery systems also exist due to its availability in a wide range of pharmaceutical dosage forms namely as gels, ointments, tablets, patches, films, and microspheres. These mucoadhesive-based dosage forms are used in drug delivery through different routes, e.g. buccal, nasal, ocular, vaginal, rectal, and GI routes. Drug delivery from the formulations occurs following bioadhesion to the biological substrate. Here the performance of mucoadhesive-based formulations is influenced by the nature of bioadhesive bond that is likely to exist in a microenvironment. Moreover, the performance is also affected by several factors that exist in the microenvironment such as pH, temperature, ionic strength, and water content will all affect bond durability [4]. It is, therefore, imperative to evaluate the bio/mucoadhesive properties of bioadhesive drug delivery systems to enable optimization and development of effective bioadhesive dosage forms.

Bioadhesive systems have been in use over the past few decades, however no standard test methods have been specifically designed for bioadhesion characterization. This places limitations upon the direct comparison of data obtained from different research groups. Despite this, researchers have continued to develop novel ways in which the adhesive potential of polymer systems may be ranked in relation to their adhesiveness [5]. Most commonly used techniques of bioadhesion characterization are based of in vitro in nature. Only few in vivo experimental set-ups have been successfully implemented to date.

2.2.1 In vitro Testing

In vitro tests are by far the most common techniques for the determination of bioadhesive strength [5]. Importantly, determination of the bioadhesive strength provides the success rate of a given bioadhesive formulation. Since no standard apparatus is available for testing bioadhesive strength, an inevitable lack of uniformity between test methods has arisen.

2.2.1.1 Methods measuring the force of detachment

Mucoadhesiveness determination through the adhesive strength between bioadhesive polymer and the attached substrate is most commonly used in vitro technique. The adhesive strength at such a bonding interface can be measured by measuring the force required to detach one entity from the other through the application of an external force. As such the destruction of the adhesive bond is usually under the application of either a shearing, tensile or peeling force [4–6]. Force of separation in bioadhesive testing is the application of force perpendicularly to the tissue/adhesive interface, during which a state of tensile stress is generated. However, during the shear stress, the direction of the forces is reoriented so that it acts along the joint interface. In both tensile and shear modes, an equal pressure is distributed over the contact area [7]. The peel test is based on the calculation of energy required to detach the patch from the substrate. The peel test is of limited use in most bioadhesive systems. However, it is of value when the bioadhesive system is formulated as a patch [8]. In tensile and shear experiments, the stress is uniformly distributed over the adhesive joint, whereas in the peel strength stress is focused at the edge of the joint. Thus tensile and shear measure the mechanical properties of the system, whereas peel strength measures the resistant of the peeling force [4].

Literature review indicates that the most common technique used for the measurement of bioadhesion has been based on tensile-based methods. In particular the use of modified textural analysers has been extensively studied. In the mid 1990s, Tobyn et al. [9] validated a new technique of mucoadhesion testing using a TA-XT2 texture analyser and porcine stomach tissue. Further work by Wong and co-workers [10] closely looked at the effect of various instrumental parameters on candidate mucoadhesive polymers. Their findings indicated that the mucoadhesive determination of polymers could be influenced by variables such as contact force; contact time and the speed of probe removal from the mucosal tissue. A longer contact time and higher probe speed were found to give a greater degree of sensitivity arising from greater reproducibility in results along with higher measurement values. In terms of the contact force, the authors determined that a certain level of force affected the mucoadhesion, beyond which further increases in force had little effect. In conclusion, the authors stated that both the work of adhesion and peak detachment force could be used in the evaluation of mucoadhesiveness of candidate adhesive polymer systems. Accordingly McCarron et al. [8,11,12] and Donnelly [13] have reported extensively on the use of TA-XT2 analyser (Fig. 2.1) operating in tensile test mode, to measure the force required to remove bioadhesive films from excised tissue in vitro. The texture analyser, operating in tensile test mode and coupled with a sliding lower platform, was also used to determine peel strength of similar formulations (Fig. 2.2) [8].

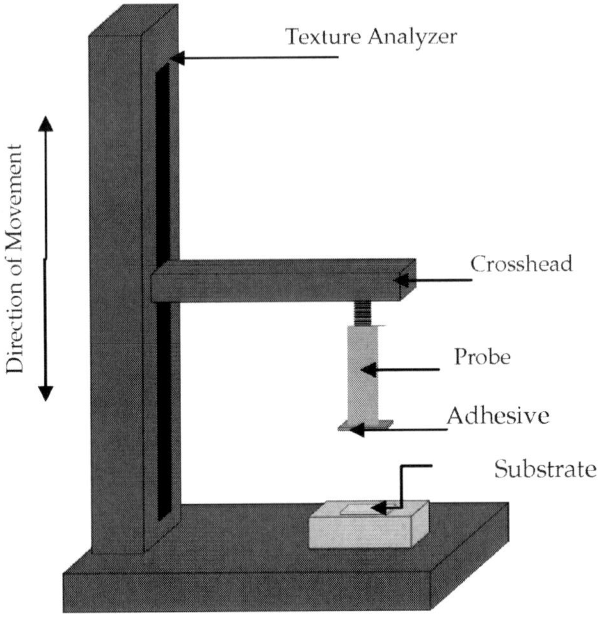

Figure 2.1 Texture profile analyser in bioadhesion test mode [4].

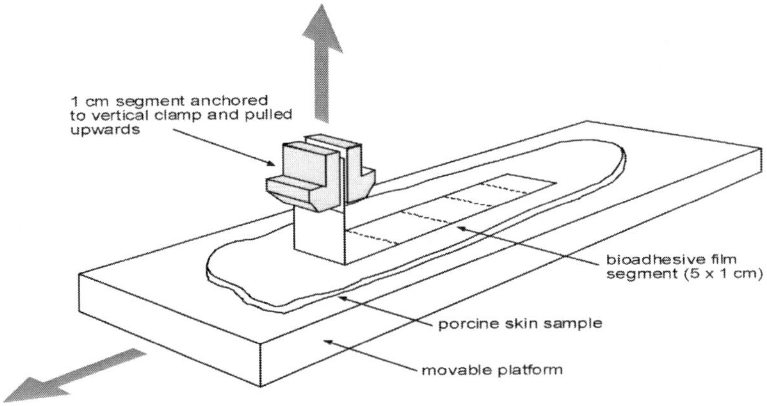

Figure 2.2 Simplified representation of a typical test set-up used to determine peel strength of bioadhesive films [4].

Mucoadhesion of a range of candidate polymers was also determined using a modified version of the Wilhelmy plate surface technique [14] (Fig. 2.3). The device consisted of a glass plate, which is coated with the polymer solution under investigation, and was suspended from a microbalance. The polymer-coated plate was then slowly dipped into a beaker of mucus. The polymers adhesiveness ranking was based upon the work required to remove the polymer-coated glass slides from the mucus solution. This technique had the advantage of allowing the analysis of mucoadhesion under different environmental conditions via simple modification of instrumental set-up. However, Mikkos and Peppas [6] pointed out the shortcomings of this technique due to the possible dissolution of the polymer upon testing. They suggested that this effect may be limited if polymer plates of the candidate material were used instead of polymer-coated glass plates. Furthermore, Wong et al. [10] also noted shortcomings of lack of biological tissue in such a set-up may not represent true mucoadhesion. Despite the simplicity and efficiency of such a technique, tensile tests provide an incomplete picture of the process of mucoadhesion. It must be borne in mind that most mucoadhesive delivery systems will tend to exhibit other mechanical forces, such as the shear stresses exhibited within the buccal cavity or GI tract [15,16]. The effect of both these important forces on measuring the adhesive bond was made possible via the use of a dual tensiometer such as the one used by Leung and Robinson [15] (Fig. 2.4).

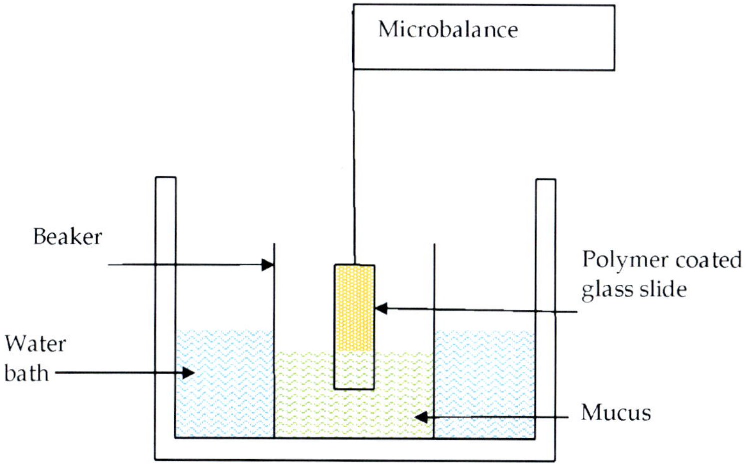

Figure 2.3 Illustrating the modified Wilhelmy plate surface technique, for mucoadhesion determination. Modified from Smart et al. [14].

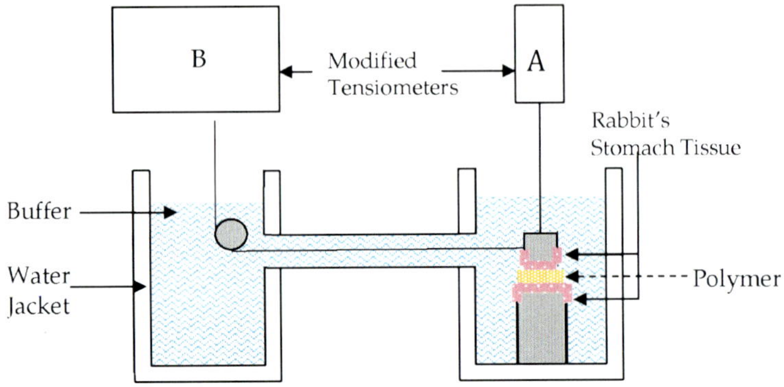

Figure 2.4 Illustrating the modified dual tensiometer used by Leung and Robinson [15].

Mortazavi and Smart [16] investigated the importance of the nature of the mucosal surface within an in vitro experiment; mucus gel, rat small intestine and a control of PVC tape (non-sticky side used) were all investigated. The authors determined the rank of adhesion as follows: PVC > Rat small intestine > mucus gel, indicating that a specific mucus/mucoadhesion interaction is not paramount in the adhesive ranking of these polymers. The apparatus used within the study had the potential to measure tensile and shear forces and was similar to the one employed by Leung and Robinson [14]. In fact the difficulty the authors had in measuring shear stresses within the study could point to why the use of such a technique has been limited to date.

Grabovac et al. [17] employed a method that does in fact allow for the measurement of mucoadhesion under the application of a shearing stress. The test is centred on the use of a rotating cylinder, wherein the adhesion time of various compressed polymer discs to porcine small intestinal mucosa was established. The procedure itself involved attaching polymer discs to freshly excised porcine intestinal mucosa which had prior been spanned onto a stainless steel cylinder. The cylinder was then placed within a dissolution apparatus and rotated at a speed of 125 rpm. Changes in the test system were visually determined every 30 min and measured until all the discs were either disintegrated or detached from the mucosal surface. The whole set-up used in this study is an USP dissolution apparatus that is typically used for testing dissolution of tablet formulations. Since this apparatus is commonly available in pharmaceutical pre-formulation testing laboratories it can be considered an economic method of bioadhesion analysis.

2.2.1.2 Rheological measurement of mucoadhesion

Rheological profiling of polymer–mucus mixtures can also provide an acceptable in vitro model that is representative of the true in vivo behaviour of a mucoadhesive polymer [18]. Hassan and Gallo [19] initially implemented rheological techniques to study the mucoadhesive properties of range of polymers. Rheological data enables examination of mucoadhesion forces of polymers by monitoring the viscosimetric changes in a system that is composed of the polymer and mucin. The energy associated with the physical and chemical bonds of the mucin-polymer interaction can be transformed into mechanical energy or work. The work causes the rearrangements of the macromolecules that in turn change the viscosity of the system [20]. Importantly, polymers with strong mucoadhesive ability are expected to have a greater viscosity in mixtures with mucins compared to the sum of polymer and mucin viscosities and this rheological synergism may be used to evaluate their bioadhesive properties. Viscosity or elasticity is experimentally determined for their mixture and the data are compared with the rheological properties of the polymer and the mucin separately. The viscosity of the mixture (η_t) can be expressed through the following relationship [21].

$$\eta_t = \eta_m + \eta_p + \eta_b, \tag{2.1}$$

where η_m and η_p are the individual viscosities of mucin and bioadhesive polymer, respectively, and η_b is a factor evolving from the interaction between mucin and polymer and is termed as bio-/mucoadhesive component. Subsequently, the bioadhesion force (F) represents the additional intermolecular frictional force per unit area and is defined by the following equation:

$$F = \eta_b \times \sigma, \tag{2.2}$$

where σ is the shear gradient (s^{-1}).

Hassan and Gallo [19] compared the rheological properties of binary polymer/mucus blends to the rheological sum of similarly concentrated mono components of mucus and polymeric systems.

Within this study bioadhesion forces (*F*) of many polymers were determined such as, poly(ethylene glycol) (4 kDa), bovine serum albumin, dextran (71.5 kDa), polybrene (4.5 kDa), cationic gelatin (193 kDa), chitosan (652 kDa), poly(aspartic acid) (35 kDa), heparin (48.5 kDa) and poly(acrylic acid) (90 kDa) in mixtures with commercial porcine gastric mucin at pH 1.0 (0.1 N HCl) and pH 5.5 (0.1 N acetate buffer). The highest *F* values were observed for chitosan and poly(acrylic acid) at pH 5.5 with polymer concentrations of 1% and 2.5% w/v, respectively. On the contrary, poly(ethylene glycol), polybrene, poly(aspartic acid) and heparin polymers showed very low mucoadhesive properties due to low molecular weights. Overall findings showed that the mucoadhesive polymer/mucus mixtures exhibited synergistic rheological profiles, the causes of which were attributed to bond formation between polymer and mucus culminating in an increase in total system structure. Since this pioneering work, there have been numerous rheological studies of polymer/mucus interactions [5]. Hassan and Gallo (1990) [19] also used simple viscometric measurements to quantify the bioadhesivness between porcine gastric mucin and polymer. Here the authors used Brookfield viscometer to measure the viscosities of mucin in absence or presence of different polymers. Viscosity readings were then used to determine the bioadhesive strength.

Madsen and co-workers [22] indicated that the rheological synergism was found to arise only within a certain polymer concentration range and that this concentration was dependent upon the polymer under investigation. Furthermore authors discussed the benefit of using non-destructive dynamic rheological analysis to measure the interaction parameters as opposed to destructive flow technique that had been previously implemented. In an extension to this work, Madsen and colleagues [23] determined the interactions between four mucoadhesive polymers (Noveon®, Pemulen® TR-2, carageenan and sodium carboxymethylcellulose) and a homogenized mucus gel. Using a dynamic rheological technique they showed that incorporation of a mucoadhesive polymer into mucus gel produced rheological behaviour that was indicative of a weakly cross-linked gel network. In addition, they noted that optimum gel strengthening occurred within a weakly acidic environment, reinforcing the idea that optimum conformation and degree of ionization of the polymer and mucus molecules are important factors for mucoadhesion. In contrast to the other measured polymer systems, the study showed that carboxymethylcellulose/mucus mixes exhibited no synergy at all [5].

Tamburic and Craig [24] work also validated that the rheological measurements can be used as a method of ranking mucoadhesive order. Within this study, the authors compared tensile mucoadhesive testing with shear rheological measurements using three polyacrylic acid-based polymers. They found that both measuring techniques gave the same rank order over adhesiveness for the polymers studied.

The isolation of fresh mucus for mucoadhesion analysis can be a laborious process. This has led many authors to turn their attention to the use of commercially available dried mucus, which is then rehydrated before analysis. Rossi et al. [25] undertook work that validated the use of such mucus material in rheological testing. Within this study, investigators determined if there was in fact any variation in mucoadhesion testing through the use of different mucus-derived substrates. The authors rheologically studied the effect of two different freeze-dried porcine mucins and one type of freeze-dried bovine mucin on rheological parameters. Through their investigations, they discovered that the choice of mucus type was important in particular for the ion- sensitive polyacrylic acids. Polyacrylic acids showed only positive interaction with the bovine-derived mucin, whereas the less ion-sensitive NaCMC had positive interaction with all tested mucus. The lack of interaction between polyacrylic acid and the porcine mucin was suggested to be due to the greater number of adhesion-impeding ions present within porcine mucin. In conclusion, authors stated that mucoadhesive polymers can be affected not only by the biochemical structure of glycoproteins but also by the way the mucin sample is prepared [5].

Kocevar-Nared and co-workers [26] also investigated the effect of mucus type on mucoadhesion testing. In this study, re-hydrated dried crude porcine gastric mucin was compared to that of freshly isolated mucus from the porcine gut. The study showed that no matter what loading of dried mucus was reconstituted; rheological properties similar to freshly isolated mucus could not be achieved. The authors concluded that the isolation procedures changed the physicochemical properties of the mucus to such an extent that reconstitution resulted on mucus of an altered structure. Despite experimental work showing that re-hydrated commercial mucus does not possess the exact viscoelastic structure of freshly isolated mucus, Hägerström et al. [27] pointed out that the use as of fresh mucus gels is not unquestionable since it requires reproducible isolation and purification procedures so that the amount of glycoproteins able to interact always remains the same.

The rheological method is an easy in vitro approach for testing the bioadhesion of polymeric drug delivery system. Rheology measures the mechanical properties of the system, i.e. the resistance against flow and deformation, assessing the changes the system undergoes in the presence of mucin. Nevertheless, rheology does not provide any direct information on what occurs at the interface, because the two phases—mucin and polymer—are mixed together prior to the experiment [20]. Furthermore, the data obtained from rheological testing is not always reliable and it is often used in relation with other techniques. And, unlike patches or films, rheological testing is applicable for testing semi-solids and liquids dosage forms, whilst rupture tensile strength test can be applied to solids and semi-solids dosage forms. Additionally, care should be taken to monitor experimental variables, e.g. hydration, time of hydration, composition, pH, and temperature that are critical for obtaining reproducible results.

2.2.1.3 Flow-through or falling liquid film method

Rango Rao and Buri (1989) [28] proposed the first flow-through method for mucoadhesion testing. As the name suggests, this method is used for testing mucoadhesion of pharmaceutical dosage forms that are administered in the regions of the human body, where mucosal tissues are continuously exposed to a flow of biological fluids. For example, in region of GIT, ocular, nasal or vaginal areas where regionally secreted biological fluids are under continuous motion over mucosal tissues. This flow of biological fluids in the microenvironment washes off a dosage form from the surface of a mucosal tissue [21]. The typical experimental set-up of this apparatus is given in Fig. 2.5. Here selected mucosal membrane (e.g. porcine stomach, intestinal and buccal mucus) is affixed onto the slide followed by application of candidate polymeric formulation onto the membrane. The slide is placed in inclined position and set-up in a thermostatically controlled environment at 37°C. A simulated body fluid is pumped at constant rate that washes off the formulation from the surface of the mucosal membrane. The extent of formulation retention is used to assess mucoadhesiveness of the given formulation. Formulation retention is tested by analysing the contents of collecting chamber for a non-adhered drug, die or a mucoadhesive polymer by using spectrophotometry (UV-Vis or fluorescence) or by using chromatographic techniques (high-performance liquid chromatography or gel permeation chromatography). Additionally, the amount remaining on the mucous membrane can be monitored using a coulter counter, in the case of particulate systems. Nielsen et al. (1998) [29] reported visualization, by means of polarized light microscopy, of formation of liquid-crystalline mesophase on the mucous membrane after the flowing of the fluids. The flow-through method can be used for evaluating mucoadhesive properties of micro-/nanoparticle and semi-solid based dosage forms, which otherwise are difficult to measure using tensile approach, e.g. with texture analyser set-up, as described in the Section 2.1.1.

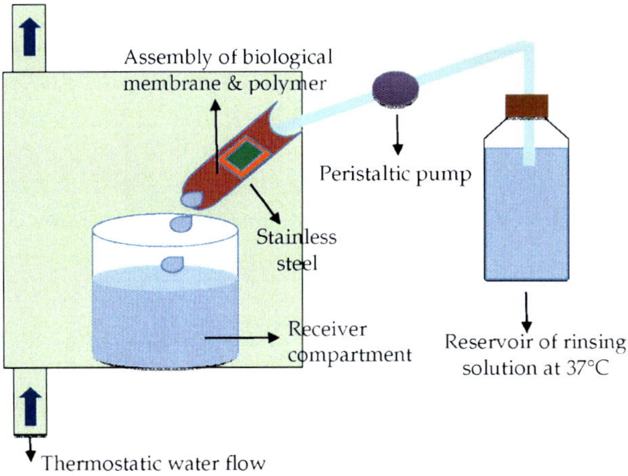

Figure 2.5 Illustration of a flow-through mucoadhesive apparatus set-up [28].

2.2.1.4 Imaging techniques

In addition to above in vitro techniques imaging has also been used as an advance and modern technique for measuring bioadhesion. Importantly atomic force microscopy (AFM) and confocal laser scanning microscopy (CLSM) are two modern techniques that offer resolution at micro- or nanometric level.

AFM or scanning force microscopy (SFM) is a very high-resolution type of scanning probe microscopy that has been used for imaging, measuring, and manipulating matter at the nanoscale (enlarges by nearly 10^9-folds) [41]. A cantilever with a sharp tip (probe) is scanned over the surface of the sample to gather the information. The deflection of the probe is measured using a laser spot reflected from the top surface of the cantilever into an array of photodiodes. Bioadhesion is measured from plotting a force-distance curve that measures the forces between the tip and the sample surface [30]. For bioadhesion determination using AFM, the probe is coated with adhesive material, which is generally spherical in shape, interacts with the surface of mucous membrane [31]. Using AFM technique Deacon et al. (2000) [32] determined the interaction between purified gastric mucin and a mucoadhesive cationic polymer, chitosan. It was concluded that the interaction was clearly electrostatic in nature, with some involvement of hydrophobic interactions. Joergensen et al. (2011) [33] assessed the mucoadhesion of the three commercially available types of pectin by AFM and SPR techniques. Image analysis using AFM showed that the low-ester substituted pectin indicated a high mucoadhesion when compared with high-ester and amidated pectin. These adhesion results were in confirmation with SPR technique.

CSLM is a technique for obtaining high-resolution optical images with depth selectivity. Confocal microscopy allows acquiring in-focus images from selected depths. Keely et al. (2005) [34] used CSLM technique to study the adhesion of a range of polymers based on poly(2-(dimethylamino-ethyl) methacrylate (pDMAEMA) with human mucus-secreting and non mucus-secreting intestinal cell monolayers. Results obtained from this study indicated better visualization, following staining with ethidium homodimer-1 (EthD1). It was also noted that the monolayers used in the study can be used to assess polymer mucoadhesion and can give similar results as that of an isolated rat intestinal sacs. Takeuchi et al. (2005) [35] examined the mucosal layer of rat intestine in order to detect liposomes by using CSLM. For this purpose, a fluorescence marker (1,1′-dioctadecyl-3,3′,3′-tetramethylindo carbocyanine perchlorate) was formulated into the liposome particles. After oral administration of the particulate system the intestinal tube was removed and cut into various sections. Each section was cryostat to a thickness of 10 μm, which was used for CSLM investigations.

Measuring the residual liposomes on the mucosa of the intestinal tube enabled assessment of the mucoadhesion profiles.

2.2.1.5 Other in vitro tests

Other adhesive testing methods that allow an alternative insight into the method and degree of mucoadhesion have also been explored. These techniques include fluorescent probe technique, surface plasmon resonance (SPR), dielectric spectroscopy, optical biosensor technique, adhesion weight method, colloidal gold staining method, viscometric method, thumb test, wash-off method and Coulter-counter method.

Park and Robinson [36] determined the effect of various polymer and mucin interactions via the use of fluorescent probes. This technique involved labelling of lipid bilayer of cultured human conjunctiva cells with the fluorescent probe pyrene. The adhesion of polymers to these cells caused a change in fluorescence due to surface compression when compared to control cells. This degree of change in fluorescence is proportional to the amount of polymer binding. Furthermore, the use of another probe also allowed the effect of polymer charge, charge sign and density on adhesion to be determined. Their findings suggested that highly charged carboxylated anions exhibit the best properties for bioadhesive drug delivery systems. Mikos and Peppas [6] highlighted the advantages of this technique that enable adhesive bond determination at molecular level. Despite this, these authors also pointed out the limitations of such a technique stating that such measurements did not take into account those macromolecules that bind perpendicularly. Batchelor and co-workers [37] further modified this technique. Within this study fluorescently labelled alginate solutions of known rheological profile were delivered onto porcine oesophageal tissue. A washing solution was applied at a specified rate to mimic saliva flow, and the eluted material collected with the degree of retention over time measured via fluorimetric analysis. Investigators showed that after 30 min up to 20% of the applied alginate dose remained associated with the tissue, regardless of the type alginate selected.

Kockisch and colleagues [38] developed a semi-quantitative image analysis-based technique for the in vitro and in vivo detections of polymers with an affinity for the mucosal surfaces of the oral cavity. This technique was used to analyse various well-known mucoadhesive polymers, allowing the visualization of the adhering polymers to buccal cell scrapings. Visualization of adhesion was aided through staining with 0.1% (w/v) of either Alcian blue (for polyanions) or Eosin (for chitosan) solution with uncomplexed dye being removed with 0.25 M sucrose washings. The extent of polymer adhesion was then quantified by measuring the relative staining intensity of control and polymer-treated cells by image analysis.

Takeuchi and co-workers [35] looked at the measurement of mucoadhesion of various adhesive polymers via the BIACORE instrumentation. This system is based on the principle underlying an optical phenomenon called SPR. Such a system measures the refractive index, which varies with the solute content of a solution that comes in contact with the sensor chip. An SRP response is achieved when a molecule becomes attached to the surface of the sensor chip as the solute concentration on the chip increases. As such, quantitative measurements can be achieved via the binding interaction between the chip surface and one or more functional groups such as NH_2, SH, CHO and COOH. The procedure itself involved the immobilization of mucoadhesive polymer on the sensor chip surface with a mucin suspension being passed over the sensor chip for a predefined time. When the mucin particle binds to the polymer on the sensor chip surface, the increased response is measured; when they dissociate, the response will fall. Such an instrument set-up allows for the real-time measurement and label-free detection of polymer mucin binding.

Dielectric spectroscopy that involves the study of the response of materials to an applied electric field can be used to derive structural information. Hägerström (2003) [39] was first to report the application of low frequency dielectric spectroscopy technique to obtain information about separate gels and mucosa, but also assessed the compatibility of intimate contact between the mucus and the gel. The compatibility is obtained as a measure of the ease with which the charged particle passes a barrier between the gel and the mucosa. A low barrier indicates a high degree of compatibility between the gel and the mucus layer and vice versa. Within this work the authors measured compatibility of different pharmaceutical gels, e.g. Carbopol 934, sodium carboxymethylcellulose, chitosan hydrochloride, and poloxamer with freshly excised porcine nasal mucosa. This technique involves the study of material response to the application of an electrical field. As shown in Fig. 2.6, the samples under investigation are placed between the cylindrical electrodes in three configurations. For each gel formulation three successive measurements are made. First, a fresh gel sample is analysed (Fig. 2.6b). Second, fresh samples of mucosa are measured (Fig. 2.6c) and, finally, a new gel sample is placed between the mucosa samples (Fig. 2.6d) and dielectric measurements are repeated. From the responses, the impedance or permittivity of the sample is obtained and the property of charges changing in the system can be determined [39].

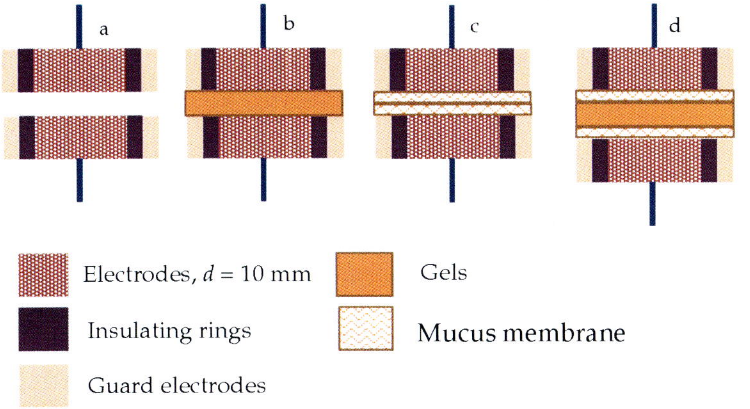

Figure 2.6 Schematic representation of apparatus for dielectric measurements [39].

Sigurdsson et al. (2006) [40] used alternative technique to evaluate molecular interactions between polymer and mucus by using the optical biosensor or resonant mirror biosensor technique. The optical biosensor technique allows monitoring interaction between two molecules in real time without the need for labelling or tagging, provided one of the molecules can be immobilized, with covalent or non-covalent bonds, on the surface of such system and the other stays in solution above the surface. Binding of molecules in solution to surface-immobilized molecules results in change of the refractive index of the medium near the surface and this change is detected by the screening of a laser beam. Using this technique Sigurdsson et al. (2006) [40] quantified interactions between mucus glycoprotein and several different polymers. In this study, bovine submaxillary mucin and non-glycosylated bovine serum albumin were also used as a substrate. Polymers such as carboxymethylcellulose sodium salt, hydroxypropyl methylcellulose, Carbopol® 934, chitosan, sodium alginate and sodium hyaluronate were used for investigations. The results of this study specified the need to clarify the mucoadhesion concept, as noted from the results the viscosity enhancing polymers do not interpenetrate mucus layers and also do not interact with glycoproteins, therefore it should not be considered as mucoadhesives. Instead the polymers showed some non-specific adhesion to mucus surface and the binding strength was dependent on the degree of swelling/hydration, which represents apparent mucoadhesives. However, this technique can be used for better understanding the physicochemical mechanism and nature of the mucin-polymer interaction and allow quick screening of mucoadhesive polymers. Further advantage of this method of analysis is the requirement of only small quantities of (≈10 mg) of polymer samples for testing.

Chowdary and Srinivasa (2003) [41] used a simple wash-off test to determine the mucoadhesive property of polymeric system. Within this test, the mucosal tissue was attached onto a glass slide with the help of a double-sided cyanoacrylate tape. Thereafter, the polymeric delivery system was applied onto the surface of the mucosal tissue. Subsequently this assembly was held vertically in an USP tablet disintegrator apparatus, which was containing physiological solution maintained at 37°C. Similar to disintegration testing of tablets, this test applies an upwards/downward movement to the mucus tissue-delivery matrix system. The time for the complete detachment of the delivery system from the mucosal surface is used to determine mucoadhesiveness of the polymeric delivery system.

Other less commonly used techniques include colloidal gold staining technique, viscometric method and thumb test. Park (1989) [42] proposed the colloidal gold staining technique to study the bioadhesion. In this method, mucin-conjugated gold particles interact with bioadhesive hydrogels that develop red colour on the surface. The degree of bioadhesiveness is detected by measuring the intensity of red colour on hydrogel surface or by measuring the drop in conjugate concentration, measured at 525 nm. Kamath and Park (1994) [43] used thumb test using physical application of thumb pressure onto the adhesive surface for defined time period and the adhesive strength was measured by determining the difficulty of pulling the thumb from the surface. However, the application of pressure, lack of numerical value and varying individual perception could lead to erroneous results by thumb testing.

To evaluate mucoadhesion of particulate system such as polymer-coated liposomes, Takeuchi et al. (1994, 1996) [44,45] used Coulter-counter method. In this method, rat intestinal tube was isolated and cut into three equal lengths of 15 cm each. Each tube was washed inside with saline solution and filled with appropriate amount of liposomal suspensions. Each side of the tube was sealed with closers and incubated in saline solution at 37°C for more than 15 min. By measuring the number of liposomal particles before and after incubation with a Coulter counter, the percentage mucoadhesive was determined using following equation:

$$\text{Adhesive}\% = \frac{(N_0 - N_s)}{N_0} \times 100, \tag{2.3}$$

where N_0 and N_s are the number of liposomes before and after incubation, respectively.

2.2.2 In vivo studies

To date, limited number of in vivo bioadhesion studies have been performed; this is due to the cost, time constraints and ethical reasons involved in in vivo testing. Therefore, in vivo mucoadhesive studies are less commonly seen in the literature than the in vitro testing. Despite these concerns, in vivo testing is still the important if the true mucoadhesive potential of a system is to be determined. As such, in vivo techniques have found their most extensive use in the analysis of potential oral mucoadhesive dosage forms.

Ch'ng et al. (1985) [46] studied the transit of various ^{51}Cr radio-labelled polyacrylic acid beads through the rat GI tract. The beads were fed to the rats and at various time intervals rats were sacrificed. The rat's intestine was then dissected into 20 equal lengths and the amount of radiation in each section measured, thus allowing the transit overtime to be realized. Such an experimental technique obviously had its limitations due to the requirement for the sacrifice of the subject. The development of a non-invasive technique to determine the transit of mucoadhesive polymers was undertaken in the work by Davis [47]. Here, the transit of polymers could be imaged via labelling of the mucoadhesive polymer system with a γ-emitting nucleotide. Detection of the polymer systems within the GI tract was then determined via γ-scintigraphy. Albrecht et al. [48] also utilized non-invasive imaging technique, magnetic resonance imaging, to localize the point of release of thiolated polymers from dosage forms via the use of gadolinium. In vivo mucoadhesion was determined by ascertaining the residence time of the fluorescently tagged thiomer on intestinal mucosa of rats after 3 h. This technique allowed the comparison of mucoadhesive properties of candidate polymer systems for oral drug delivery in vivo.

2.3 Conclusion

Bio- or mucoadhesive dosage forms offers localized drug delivery at the site of application. This is due to their attachment and prolonged retention at the site of application, which in turn is dependent upon the degree of bioadhesiveness of the dosage forms. Therefore, it becomes integral part of formulation development process to characterize the relative bioadhesiveness of different polymeric-based drug delivery systems. By using the above-mentioned in vitro and/or in vivo techniques the bioadhesive strength of a given dosage form can be determined. To a large extent in vitro techniques could offer early insights of bioadhesiveness, however careful selection of a given method is necessary to get reliable results. Selection of a given in vitro method is also dependent upon the type of dosage forms. In some instances multiplicity of methods could be applied to obtained reliable information. For example, modern techniques such as AFM and CSLM could offer better understanding of mucus–polymer interaction that can also be confirmed by running parallel SPR or staining techniques. However, more work is required to assess the mucus–polymer interaction in in vivo setting, as only few studies have been reported so far in the literature.

References

1. Kinloch, A. J. (1980). The science of adhesion, *J. Mater. Sci.*, **15**, 2141–2166.

2. Henriksen, I., Green, K., Smart, J. G., Smistad, J. Karlsen. (1996). Bioadhesion of hydrated chitosans: An in vitro and in vivo study, *Int. J. Pharm.*, **145**, 231–240.
3. Woodley, J. (2001). Bioadhesion: New possibilities for drug administration? *Clin. Pharmacokinet.*, **40**, 77–84.
4. Shaikh, R., Raj Singh, T. R., Garland, M. J., Woolfson, A. D., Donnelly, R. F. (2011). Mucoadhesive drug delivery systems, *J. Pharm. Bioallied Sci.*, **3**(1), 89–100.
5. Andrews, G. P., Laverty, T. P., Jones, D. S. (2009). Mucoadhesive polymeric platforms for controlled drug delivery, *Eur. J. Pharm. Biopharm.*, **71**(3), 505–518.
6. Mikos, A. G., Peppas, N. A. (1986). Systems for controlled release of drugs. V: Bioadhesive systems, *STP Pharma. Sci.*, **19**, 705–715.
7. Park, K., Park, H. (1990). *Test Methods of Bioadhesion, Bioadhesive Drug Delivery Systems.* In: Lenaerts V, Gurney R, ed. Florida, Boca Raton: CRC Press.
8. McCarron, P. A., Donnelly, R. F., Zawislak, A., Woolfson, A. D., Price, J. H., McClelland, R. (2005). Evaluation of a water-soluble bioadhesive patch for photodynamic therapy of vulval lesions, *Int. J. Pharm.*, **293**, 11–23.
9. Tobyn, M. J., Johnson, J. R., Dettmar, P. W. (1995). Factors affecting in vitro gastric mucoadhesion. I: Test conditions and instrumental parameters, *Eur. J. Pharm. Biopharm.*, **41**, 235–241.
10. Wong, C., Yuen, K., Peh, K. (1999). An in-vitro method for buccal adhesion studies: Importance of instrument variables, *Int. J. Pharm.*, **180**, 47–57.
11. McCarron, P. A., Woolfson, A. D., Donnelly, R. F., Andrews, G. P., Zawislak, A., Price, J. H. (2004). Influence of plasticiser type and storage conditions on the properties of poly(methyl vinyl ether-co-maleic anhydride) bioadhesive films. *J. Appl. Polym. Sci.*, **91**, 1576–1589.
12. McCarron, P. A., Donnelly, R. F., Zawislak, A., Woolfson, A. D. (2006). Design and evaluation of a water-soluble bioadhesive patch formulation for cutaneous delivery of 5-aminolevulinic acid to superficial neoplastic lesions, *Eur. J. Pharm. Sci.*, **27**, 268–279.
13. Donnelly, R. F., McCarron, P. A., Zawislak, A. A., Woolfson A. D. (2006). Design and physicochemical characterisation of a bioadhesive patch for dose-controlled topical delivery of imiquimod, *Int. J. Pharm.*, **307**, 318–325.

14. Smart, J. D., Kellaway, I. W., Worthington, H. E. (1984). An in-vitro investigation of mucosa-adhesive materials for use in controlled drug delivery, *J. Pharm. Pharmacol.*, **36**, 295–299.
15. Leung, S. H., Robinson, J. R. (1988). The contribution of anionic polymer structural features to mucoadhesion, *J. Control. Release*, **5**, 223–231.
16. Mortazavi, S. A., Smart, J. D. (1995). An investigation of some factors influencing the in vitro assessment of mucoadhesion, *Int. J. Pharm.*, **116**, 223–230.
17. Grabovac, V., Guggi, D., Bernkop-Schnürch, A. (2005). Comparison of the mucoadhesive properties of various polymers, *Adv. Drug Deliv. Rev.*, **57**, 1713–1723.
18. Riley, R., Smart, J., Tsibouklis, J., Dettmar, P., Hampson, F., Davis, J. A., Kelly, G., Wilber, W. (2001). An investigation of mucus/polymer rheological synergism using synthesised and characterised poly (acrylic acid)s, *Int. J. Pharm.*, **217**, 87–100.
19. Hassan, E. E., Gallo, J. M. (1990). A simple rheological method for the in vitro assessment of mucin–polymer bioadhesive bond strength, *Pharm. Res.*, **7**, 491–495.
20. Carvalho, F. C., Bruschi, M. L., Evangelista, R. C., Gremião, M. P. D. (2010). Mucoadhesive drug delivery systems, *Braz. J. Pharm. Sci.*, **46**(1).
21. Khutoryanskiy, V. V. (2011). Advances in Mucoadhesion and Mucoadhesive Polymers, *Macromol. Biosci.*, **11**, 748–764.
22. Madsen, F., Eberth, K., Smart, J. (1998). A rheological examination of the mucoadhesive/mucus interaction: The effect of mucoadhesive type and concentration, *J. Control. Release*, **50**, 167–178.
23. Madsen, F., Eberth, K., Smart, J. (1998). A rheological assessment of the nature of interactions between mucoadhesive polymers and a homogenised mucus gel, *Biomaterials*, **19**, 1083–1092.
24. Tamburic, S., Craig, D. M. (1997). A comparison of different in vitro methods for measuring mucoadhesive performance, *Eur. J. Pharm. Biopharm.*, **44**, 1159–167.
25. Rossi, S., Bonferoni, M. C., Lippoli, G., Bertoni, M., Ferrari, F., Caramella, C., Conte, U. (1995). Influence of mucin type on polymer–mucin rheological interactions, *Biomaterials*, **16**, 1073–1079.
26. Kocevar-Nared, J., Kristl, J., Smid-Korbar, J. (1997). Comparative rheological investigation of crude gastric mucin and natural gastric mucus, *Biomaterials*, **18**, 677–681.

27. Hagerstrom, H., Paulsson, M., Edsman, K. (2000). Evaluation of mucoadhesion for two polyelectrolyte gels in simulated physiological conditions using a rheological method, *Eur. J. Pharm. Sci.*, **9**, 301–309.

28. Rango Rao, K. V., Buri, P. (1989). A novel in situ method to test polymers and coated microparticles for bioadhesion, *Int. J. Pharm.*, **52**(3), 265–270.

29. Nielsen, L. S., Schubert, L., Hansen, J. (1998). Bioadhesive drug delivery systems. I. Characterization of mucoadhesive properties of systems based on glyceryl mono-oleate and glyceryl monolinoleate, *Eur. J. Pharm. Sci.*, **6** (3), 231–239.

30. Mathiowitz, E., Chickering, D. E., Lehr, C. M. (eds.) (1999). Bioadhesive drug delivery systems: Fundamentals, novel approaches, and development. *Drugs and the Pharmaceutical Sciences*. New York: Marcel Dekker. 696 p.

31. Cleary, J., Bromberg, L., Magner, E. (2004). Adhesion of polyether-modified poly(acrylic acid) to mucin, *Langmuir*, **20**(22), 9755–9762.

32. Deacon, M. P., McGurk, S. Roberts, C. J., Williams, P. M., Tendler, S. J., Davies, M. C., Davis, S. S., Harding, S. E. (2000). Atomic force microscopy of gastric mucin and chitosan mucoadhesive systems. *Biochem J.*, **348**, 557–563.

33. Joergensen, L., Klösgen,B., Simonsen, A.C., Borch, J., Hagesaether. E. (2011). New insights into the mucoadhesion of pectins by AFM roughness parameters in combination with SPR, *Int. J. Pharm.*, **411**(1–2), 162–168.

34. Keely, S., Rullay, A., Wilson C., Carmichael, A., Carrington, S., Corfield, A., Corfield, A., Haddleton, D. M., Brayden, D. J. (2005). In vitro and ex vivo intestinal tissue models to measure mucoadhesion of poly (methacrylate) and N-trimethylated chitosan polymers, *Pharm. Res.*, **22**(1), 38–49.

35. Takeuchi, H., Thongborisute, J., Matsui, Y., Sugihara, H., Yamamoto, H., Kawashima, Y. (2005). Novel mucoadhesion tests for polymers and polymer-coated particles to design optimal mucoadhesive drug delivery systems, *Adv. Drug Deliv. Rev.*, **57**, 1583–1594.

36. Park, K., Robinson, J. R. (1984). Bioadhesive polymers as platforms for oral-controlled drug delivery: Method to study bioadhesion, *Int. J. Pharm.*, **19**, 107–127.

37. Batchelor, H. K., Banning, D., Dettmar, P. W., Hampson, F. C., Jolliffe, I. G., Craig, D. Q. M. (2002). An in vitro mucosal model for prediction of the bioadhesion of alginate solutions to the oesophagus, *Int. J. Pharm.*, **238**, 123–132.

38. Kockisch, S., Rees, G., Young, S., Tsibouklis, J., Smart, J. (2001). A direct-staining method to evaluate the mucoadhesion of polymers from aqueous dispersion, *J. Control. Release*, **77**, 1–6.

39. Hagerstrom, H., Edsman, K., Stromme, M. (2003). Low-frequency dielectric spectroscopy as a tool for studying the compatibility between pharmaceutical gels and mucus tissue, *J. Pharm. Sci.*, **92**(9), 1869–1881.

40. Sigurdsson, H. H., Loftsson, T., Lehr, C. M. (2006) Assessment of mucoadhesion by a resonant mirror biosensor, *Int. J. Pharm.*, **325**(1), 75–81.

41. Chowdary, K. P. R., Srinivasa, R. Y. (2003). Design and in vitro and in vivo evaluation of mucoadhesive microcapsules of glipizide for oral controlled release: A technical note, *AAPS Pharm. Sci. Tech.*, **4**(3), 87–92.

42. Park, K. (1989). A new approach to study mucoadhesion: Colloidal gold staining, *Int. J. Pharm.*, **53**, 209–217.

43. Kamath, K. R., Park, K. (eds.) (1994). Mucosal Adhesive Preparations, In: Swarbrick, J., and Boylan, J. C., *Encyclopedia of Pharmaceutical Technology*, vol. 10, Marcel Dekker, New York, pp. 133–163.

44. Takeuchi, H., Yamamoto, T., Niwa, T., Hino, Y. (1994). Kawashima, Mucoadhesion of polymer-coated liposomes to rat intestine in vitro, *Chem. Pharm. Bull.*, **42**, 1954–1956.

45. Takeuchi, H., Yamamoto, H., Niwa, T., Hino, T., Kawashima, Y. (1996). Enteral absorption of insulin in rats from mucoadhesive chitosan-coated liposomes, *Pharm. Res.*, **13,** 896–901.

46. Ch'ng, H. S., Park, P., Kelly, J. R. (1985). Robinson, Bioadhesive polymers as platforms for oral controlled drug delivery. II: Synthesis and evaluation of some swelling, water-insoluble bioadhesive polymers, *J. Pharm. Sci.*, **74**, 399–405.

47. Davis, S. S. (1985). The design, evaluation of controlled release systems for the gastrointestinal tract, *J. Control. Release*, **2**, 27–38.

48. Albrecht, K., Greindl, M., Kremser, C., Wolf, C., Debbage, P., Bernkop-Schnürch, A. (2006). Comparative in vivo mucoadhesion studies of thiomer formulations using magnetic resonance imaging and fluorescence detection, *J. Control. Release*, **115**, 78–84.

Section II

Natural Adhesives

Chapter 3

Mussel Adhesives

Hongbo Zeng, Qingye Lu, Bin Yan, Jun Huang, Lin Li, and Zhi Liao

Department of Chemical and Materials Engineering,
University of Alberta, Edmonton, AB T6G 2V4, Canada

hongbo.zeng@ualberta.ca

This chapter describes recent advances on bioadhesives in the mussel byssus system and unveils the associated molecular interaction mechanisms. These studies are of both fundamental and practical importance for the design and development of novel biomimetic materials with biomedical and industrial applications.

3.1 Introduction

The physiological environment of a human body shows several common features with the wind- and wave-swept seacoasts, which experience both continuous mechanical, chemical, and biological stresses from the salinity, humidity and dramatic environmental variations such as temperature, pH, flow pattern (tides, turbulence for seashore and heart-pumped flow in human body), degradation and attacks (microbial predators in seawater, endogenous enzymes, and immunogenic attacks in human body). Interestingly, mussels are able to survive in such harsh marine environment by

Bioadhesion and Biomimetics: From Nature to Applications
Edited by Havazelet Bianco-Peled and Maya Davidovich-Pinhas

ISBN 978-981-4463-98-0 (Hardcover), 978-981-4463-99-7 (eBook)
www.panstanford.com

producing the proteinaceous byssus to adhere tightly to various substrata underwater, which has received much attention due to their unique underwater adhesion capability for the development of novel biomimetic biomaterials (i.e., coatings, adhesives, drug delivery carriers) for biomedical applications [1–5].

Since the original studies by Brown in early 1950s [6], significant progress has been achieved in the research on mussel adhesion and bio-inspired innovations in biotechnology. Different techniques have been applied to help understand the fundamental interaction mechanisms of mussel adhesion, such as in situ resonance Raman spectroscopy [7], attenuated total internal reflection Fourier transform infrared spectroscopy (ATR-FTIR) [8], quartz crystal microbalance with dissipation (QCM-D) [9], atomic force microscopy (AFM) [10–14] and surface forces apparatus (SFA) [1]. Nanomechanical techniques such as AFM and SFA can directly probe the interaction forces between different molecules and surfaces in complex vapor and solution conditions, which can provide important information toward a better understanding of the mussel adhesion mechanisms.

This chapter reviews the recent advances in the mussel adhesive system and deals with the adhesion and cohesion mechanisms of various mussel proteins in complex biological conditions as directly probed by SFA and AFM for the development of bio-inspired underwater adhesives.

3.2 Mussel Adhesive System

3.2.1 Mussel Byssus

Mussel is notorious for its ability to foul various natural and man-made surfaces such as rocks, wood structure, metals, concrete, polyvinylchloride (PVC), polymethylmethacrylate (PMMA), polytetrafluoroethylene (PTFE), other polymers, and marine organisms [15–17]. The versatile wet adhesion of mussel has attracted much attention to reveal the associated molecular structures and molecular interaction mechanisms.

The secure attachment of mussel to solid surfaces is critically controlled by the mussel byssus. Mussel byssus is essentially a bundle of threads tipped with adhesive plaques and consists of four main parts: adhesive plaque, proteinaceous threads, stem, and root (at the base of the mussel foot) (Fig. 3.1a) [1]. Byssal threads are attached to the stem of mussel foot which connects to 12 byssal retractor muscles in the shell. Byssal thread formation starts from the emerging and "tiptoeing" exploration of the mussel foot on the substratum. When a suitable patch of substratum is located, mussel foot starts a so-called surface preparation process followed by attachment. Then certain kinds of liquid proteins are secreted in the ventral groove and react to form a new byssal thread and byssal plaque. Finally, the foot initiates the byssal formation process again by locating a new spot or retracts to the mussel shell, leaving the thread connecting the mussel shell and the byssal plaque which holds fast to the substratum [18,19].

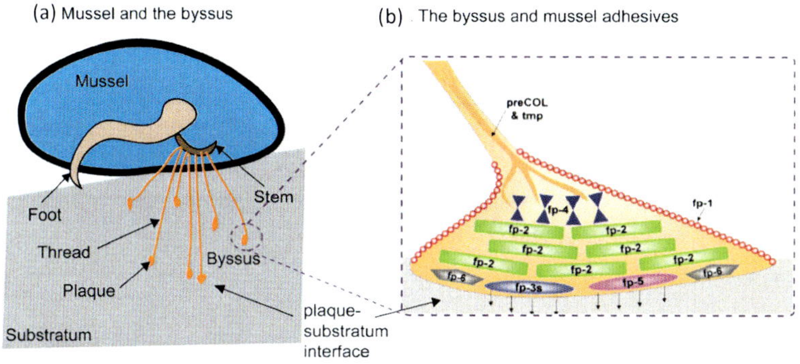

Figure 3.1 Schematic of mussel adhesive system: (a) mussel and the byssus, showing typical adult mussel with a byssus attached to a substrate, which contains four parts: plaque, thread, stem, and root (at the base of the mussel foot); (b) the byssus and mussel adhesives, showing typical byssal plaque proteins of *Mytilus* in the byssus. Panel b was modified based on a figure published in the reference [21]. Copyright © the American Society for Biochemistry and Molecular Biology.

Each mussel byssal thread has an inner core composed of collagen proteins and is coated by a natural thin layer of protective cuticle (2–5 μm thick) [1,2,7,20,21]. The structure and properties of cuticle have attracted much research attention as a new paradigm for the development of wear and damage resistant medical coatings due to its co-existence of high extensibility, stiffness, and hardness. Ultrastructural analyses of the cuticle reveal a composite structure with granules (~1 μm) dispersed in a continuous and homogeneous matrix [1,2,7]. Dopamine-metal complexes are also believed to contribute to the integrity of composite cuticles during deformation under tension [7,22].

The mussel adhesion to various substrata is mainly controlled by the properties of the adhesive plaques and the interactions at the plaque–substratum interface under complex environment conditions. The byssal plaques have a high porosity (~40%) with unique protein structure network (Fig. 3.1b), which plays a critical role in its underwater adhesion and is introduced in the following section.

3.2.2 Biochemistry of Mussel Byssus

There are roughly 25–30 different proteins in *Mytilus* byssus, and the plaque contains mussel foot protein-1 (mfp-1), -2 (mfp-2), -3 (mfp-3),-4 (mfp-4), -5 (mfp-5), -6 (mfp-6), the prepolymerized collagens (preCOL), and thread matrix proteins (tmp) which form a network shown in Fig. 3.1b. These proteins mostly contain the post-translationally modified amino acid 3,4-dihydroxyphenylalanine (DOPA) and have basic isoelectric points (pI), each of which plays an important role in the adhesion of mussel byssus. Some selected properties (e.g., molecular weight, pI, DOPA content) of these mfp proteins are shown in Table 3.1 [1,62].

Mfp-1 is a large (~110 kDa in *M. edulis* [23]), basic protein (pI = 10.5) and it is the single protein found exclusively in the cuticle of byssal thread and plaque, consisting of tandemly repeated decapeptide. The decapeptide sequence may vary with the mussel species, and the structure is AKPSY*O*OTY*K for *M. edulis* where O, O*, Y* represent trans-4-hydroxyproline, trans-2,3-*cis*-3,4-dihydroxy proline and DOPA, respectively [24,25]. DOPA content of mfp-1 is relatively high (about 10–15 mol%), which makes the protein prone to oxidation.

Mfp-2 is the most abundant protein in adhesive plaque which contributes 25–40% of the plaque protein as an important structural component [26]. Mfp-2 is present in the inner part of the plaque with a molecular weight of around 45 kDa [27,28] and a relatively low DOPA content of 5 mol%. Mfp-2 consists of 11 tandem repeats of epidermal growth factor (EGF)-like motifs, which resemble a knot structure and are stabilized by three disulfide bonds [29].

Table 3.1 Properties of mussel proteins in or near the plaque

Protein	Mass (kDa)	pI	Post-translational modifications DOPA (mol%)	Phosphorylation (mol%)	Cysteine (mol%)	Location
Mefp-1	108	10.5	10–15	—	—	Cuticle
Mgfp-2	45	9.5	5	—	—	In the plaque
Mefp-3	6	8–10	20	—	—	At the plaque–substratum interface
Mcfp-4	93	10.5	2	—	—	In the plaque
Mefp-5	9	9	28	9.4	—	At the plaque–substratum interface
Mcfp-6	11.6	9.5	3	—	11	At the plaque–substratum interface

Note: Mefp, mgfp, and mcfp are from *M. edulis*, *M. galloprovincialis*, and *M. californianus*, respectively.

Mfp-3s have the most variants (over 20) in all plaque proteins and locate at the plaque–substratum interface. Mfp-3s are the smallest adhesive plaque protein (~6 kDa) with similar pI as mfp-1 [30,31]. Mfp-3s contain 20–25 mol% DOPA, considerably higher than that of mfp-1, mfp-2 and mfp-4. Due to the small size and the presence of arginine, mfp-3s have the structural flexibility, which enables DOPA to develop cross-links and 4-hydroxy-L-argine to form hydrogen bonds during the interactions with substrata [32,33].

Mfp-4 is a relatively large protein (93 kDa) with low level of DOPA (<5 mol%) which are confined to its N and C termini [34]. Mfp-4 connects Mfp-2 and Precol-D, acting as a coupling agent between the byssal plate and thread [31]. Besides the coupling function, the understanding of the specific structure–functionality relation of Mfp-4 still remains limited.

Mfp-5 contains the highest DOPA content (~30 mol%) among all the byssal adhesive proteins, which has a molecular mass of ~9 kDa and a pI of ~9 [35]. Similar to mfp-3s, it is located at the plaque–substratum interface. The fact that both mfp-3s and mfp-5 at plaque–substratum interfaces have high DOPA content implies the important role of DOPA in mussel adhesion.

Mfp-6 has relatively low molecular weight (~11.6 kDa [36]) and DOPA content (<3 mol%) but contains ~11 mol% of cysteine. Mfp-6 is located close to the byssal plaque surface, and one possible role of Mfp-6 is providing a link between DOPA-rich surface and the bulk foot proteins [36,37]. Another important property of mfp-6 is that it can effectively reduce dopaquinones back to DOPA to maintain the redox balance for efficient mfp-3s and mfp-5 adhesion [38].

PreCOLs consist of three main kinds of collagenous proteins: preCOL-D, preCOL-P and preCOL-NG. The distinguishing letters represent their distribution in the byssal thread, with D for distal, P for proximal and NG for nongraded. All these collagen proteins contain repeat sequence $(Gly\text{-}X\text{-}Y)_n$ in triple helical configuration, with X and Y representing proline and hydroxyproline, respectively [39]. PreCOL-D provides stiffness and strength to the distal region of the byssal thread. PreCOL-P endows the proximal region of the byssal thread with elasticity and extensibility, which attaches to the retractor muscle. PreCOL-NG is believed to mediate the interaction between preCOL-D and preCOL-P [40].

3.3 Understanding Interactions of Mussel Foot Proteins

3.3.1 Techniques for Studying Molecular Interactions of Mussel Foot Proteins

As stated above, mussel proteins play the key roles in mussel adhesion and secure attachment to solid substrata in a wet and turbulent environment. It is of both fundamental and practical significance to understand the molecular interaction mechanisms of mussel proteins. Many techniques have been applied in the field, such as QCM-D, ATR-FTIR, in situ resonance Raman spectroscopy, AFM and SFA. Here we provide a general overview of these techniques for studying the mussel protein interactions and particularly focus on the indispensable force probe techniques (i.e., AFM and SFA).

3.3.1.1 General surface analytical techniques

Many surface analytical techniques can be applied to study the adsorption and molecular interactions of mussel foot proteins, such as QCM-D, Raman spectroscopy, and ATR-FTIR.

QCM-D provides sensitive mass changes ($ng \cdot cm^{-2}$ range) in real time and kinetics information for micro-scale adhesion and surface adsorption by simultaneously measuring the resonant frequency change (Δf, related to mass uptake) and energy dissipation (ΔD, related to viscoelastic properties). The adsorption of recombinant *M. gallopro-vincialis* foot protein (mgfp-5) on gold surface was quantified by employing this technique [11]. The QCM-D technique has also been used to study the viscoelastic properties and the curing or cross-linking phenomena of the blue mussel (*M. edulis*) foot protein 1, mefp-1, adsorbed on SiO_2 or hydrophobized gold surfaces [41]. Similar to QCM-D, optical techniques such as ellipsometry and surface plasmon resonance (SPR) can monitor the film thickness change during protein adsorption based on the difference in refractive index between the adsorbed protein molecules and water displaced by the proteins upon adsorption [9].

Raman spectroscopy provides important information on chemical structures, crystallinity, and molecular interactions of the materials based on Raman scattering of monochromatic light. Resonance Raman spectroscopy demonstrated that mefp-1 is able to bind Fe^{3+} ions via DOPA ligands. The chromophores are ferric catecholate complexes, and there are two different forms of the iron–protein complex in mefp-1 (the pink one at 500 nm and the purple one at 548 nm) [42]. The in situ resonance Raman spectroscopy also shows that the cuticle is a polymeric scaffold stabilized by catechol-iron chelate complexes with an unusual clustered distribution [7]. The precise localization of metal–protein complexes in mussel byssus can be determined by the use of Raman microscopy [7]. The binding of mcfp-1 to TiO_2 via bidentate coordination complexes has also been identified by resonance Raman spectroscopy [43].

ATR-FTIR was applied to probe the kinetics of appearance of mefp proteins, the interactions of mfp family, or interactions of protein to other chemicals at the surface. Previous study on alginate adsorption to germanium (Ge) and polystyrene substrates coated with mefp-1 layers shows that lysine plays a dominant role in the interaction between alginate and mefp-1 and alginate adsorption can be used as a quantitative probe for lysine in mefp-1 using ATR-FTIR [44]. It was reported that mefp-1 loses the adsorption ability to alginate after oxidized by sodium periodate [45]. The adsorption study of mefp-1 or mefp-2 on Ge surface, and mefp-2 on nonoxidized or oxidized mefp-1 adlayer by ATR-FTIR shows that mefp-1 excludes mefp-2 from the surface while oxidation of mefp-1 adlayer by sodium periodate recovers mefp-2 adsorption to the surface [46]. Thus, it was proposed that mefp-1 serves as an adhesive to the surface, and interacts with mefp-2 only after oxidative activation of DOPA in mefp-1 [46].

Other techniques such as circular dichroism (CD), small angle X-Ray scattering (SAXS), and electron paramagnetic resonance (EPR) spectroscopy can also be employed to study the structure change during protein interaction. CD was employed to detect the protein conformational change before and after adding excess periodate [47]. SAXS experiments were performed to study the structure of mefp-1 aggregates treated with $NaIO_4$ and Cu^{2+} [48]. EPR has been applied to probe the coordination environment and interaction between metal ions and proteins, such as the interaction of ferric ions with mefp-1 [42], the coordination environment of iron in mussel [49].

It is noted that in most cases several complementary techniques were applied for probing the molecular interactions of mussel proteins. Hook et al. measured the thickness, shear elastic modulus, and shear viscosity of the mussel adhesive protein film by using the QCM-D, SPR and Ellipsometry techniques [9]. The adsorption and subsequent polymerization of the mussel adhesive protein mefp-1 were monitored in an aqueous environment where the viscoelastic properties could be varied in situ by biochemical methods [9]. The interaction of ferric ions and mefp-1 was investigated by Steven et al. using resonance Raman spectroscopy, spectrophotometric titrations, and EPR [42].

3.3.1.2 Atomic force microscopy

AFM is a very sensitive technique for both qualitative and quantitative analysis of various material surfaces with nanoscale spatial resolution. The basic components of a typical AFM include cantilever with tip, piezoelectric tube (PZT) scanner, laser diode, photodiode detector, and feedback control circuit. Samples for AFM study normally do not require any special pretreatment that may alter the sample properties. AFM has been widely applied for imaging surface topography as well as molecular force measurements with typical force sensitivity of 10–100 pN, which makes single molecular force measurement feasible. It is noted that for soft materials such as proteins, it is normally difficult for AFM to obtain the absolute separation distance, contact area and surface deformation during force measurements. AFM has been used for probing mussel adhesion in several configurations: (1) a colloid probe functionalized with mfps [10,11], (2) a chemically modified AFM tip tethered with single DOPA residues to isolate the DOPA contribution in adhesion [12,13], and (3) a tipless AFM cantilever against adhesive pillar arrays inspired by mussels and geckos [14]. Lee et al. employed chemically modified Si_3N_4 AFM cantilevers and investigated the interaction of single DOPA residues with different substrate surfaces and the oxidation dependent adhesive properties of DOPA at the single-molecule level [12]. Strong and reversible adhesion was measured between a single DOPA residue and a wet metal oxide surface [12]. The adhesion of a so-called "geckel" adhesive (viz. nanostructured polymer pillar arrays) inspired from both gecko and mussel was measured using a tipless AFM cantilever coupled with an optical microscope, and it was found that the wet adhesion of the patterned surface coated with mussel-mimetic polymer was enhanced by up to 15 times [14].

3.3.1.3 Surface forces apparatus

Surface forces apparatus has been widely used for direct measurement of intermolecular and surface forces for numerous biological and non-biological systems in vapors and liquids, such as van der Waals forces, electrostatic forces, adhesion forces, hydrophobic interactions, steric interactions, specific biological interactions, friction and lubrication forces, with angstrom resolution for separation distance and nanonewton force precision [50–56]. The SFA technique has provided significant mechanistic insights into protein–substrate and protein–protein interaction mechanisms of the mussel byssal plaques.

A schematic drawing of typical SFA experiment setup is shown in Fig. 3.2. The main body of SFA includes micrometers, main stage with the single-cantilever spring (for SFA 2000), lower disk holder, upper disk holder with a piezoelectric tube, and a fully sealable chamber (Fig. 3.2a). The separation distance between two interacting surfaces can be controlled by differential micrometers over a range of ~2 mm with ~50 nm precision, by a differential spring mechanism over a range of >10 μm with a position accuracy of 0.2 nm, or by a piezoelectric tube over a range of 1 μm with <0.1 nm precision.

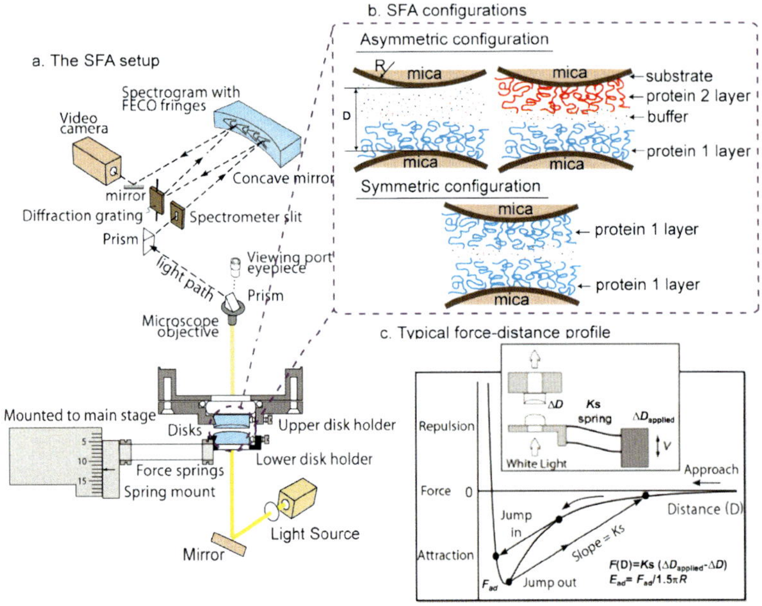

Figure 3.2 Surface forces apparatus (SFA): (a) schematic of SFA experiment setup; (b) schematic of three typical configurations for probing mussel protein interactions, i.e., asymmetric (protein 1 layer *vs.* a substrate; protein 1 layer *vs.* protein 2 layer), symmetric (two layers of the same protein); (c) schematic of typical force-distance profile showing two adhesive surfaces "jump in" during approach and "jump out" during separation.

In a typical SFA experiment, two back silvered mica sheets are glued onto two cylindrical silica discs of radius *R* and are mounted into the main chamber in a crossed-cylinder configuration (which is equivalent to a sphere of the same radius on a flat surface). Mica is commonly used as supporting substrate in SFA experiments due to its molecularly smooth feature. Other materials such as SiO_2, Au, and polymers have also been used as supporting substrata. The surface chemistry of mica can be readily modified by various physical or chemical methods, such as spin coating, self-assembly, vapor deposition, Langmuir–Blodgett deposition, and surface-initiated polymerization. Three typical configurations can be employed to probe the molecular interactions of proteins (Fig. 3.2b): a protein film against a substrate (asymmetric configuration), a protein film against a different protein film (asymmetric), or a protein film against another film of the same protein (symmetric) (Fig. 3.2b). Surface forces apparatus is also able to probe the adsorption of proteins from bulk solution to different supporting substrata in situ and in real time.

Figure 3.2c shows the illustration of typical force-distance curves obtained by the SFA. During the measurements, the lower surface supported by a "force spring" with a spring constant of Ks is moved by a distance $\Delta D_{\text{applied}}$ toward the fixed top surface, and the actual distance between two surfaces, ΔD, is determined through multiple beam interferometry (MBI) by employing fringes of equal chromatic order (FECO) [57]. The normal forces between the surfaces in various vapors or liquid media can be obtained and given by $F(D) = ks(\Delta D_{\text{applied}} - \Delta D)$ as a function of absolute separation distance D between the two surfaces. For deformable soft materials, the adhesion energy per unit area (E_{ad}) can be correlated to the adhesion force F_{ad} by $E_{\text{ad}} = F_{\text{ad}}/1.5\ \pi R$ [58]. The SFA can also provide information on so-called "hard-wall" distance, viz., the confined film thickness between the two surfaces which does not appear to significantly change with increasing the normal (compressive) load or pressure [59]. The confined protein thickness depends on the properties of protein molecules and substrata (e.g., molecular weight, rigidity of chains, hydrophobicity) and adsorption states (e.g., grafting density, conformation) which can be affected by the solution environment [47].

Several important parameters can be modulated in SFA to allow measurements in simulated biological environment for mussel foot protein interaction, such as ion type and ionic strength, pH, and temperature. Both the ion type and ionic strength may significantly affect the protein interactions [60]. Recent SFA experiments by Zeng et al. showed that divalent and trivalent cations can induce considerable adhesion of non-adhering mfps such as mfp-2 and mfp-1 [1,21,59]. Interactions between mfp-2s are mainly Ca^{2+} and Fe^{3+} mediated [21]. Solution pH is another important parameter for modulating mussel protein adhesion. Recent SFA studies have investigated the impact of pH on the adhesion of mfps, particularly mfp-3 [38]. It was found that the adhesion between two mfp-3 films at pH 5.5 and pH 7.5 could be reduced by ~80% and ~95%, respectively, as relative to that at pH 3 due to the oxidation of DOPA to dopaquinone [38]. SFA measurements show that DOPA redox plays a very important role in adhesion and cohesion of proteins in mussel plaques, and that the adhesion of mfp-3 to mica decreases with DOPA oxidation which can be recovered by the addition of antioxidants that reduce dopaquinone back to DOPA [38,47]. The experimental temperature can be readily controlled in SFA measurements, and it was found that increasing solution temperature leads to changes of the mussel protein conformations and their adhesion consequently [61].

3.3.2 Protein–Substratum Interactions

Protein–substratum interactions are dependent on the properties of protein, chemistry of the substratum surfaces and the micro-environment at the protein–substratum interface. Previous studies (mainly from force measurements by AFM and SFA) have shown that the modified prominent amino acid in mfps, DOPA, plays a critical role in mussel adhesion to various substrata, via diverse interactions including hydrogen bonding, metal coordination, cation-π interactions, and covalent bonding. DOPA-rich proteins mfp-3 and mfp-5 play a major role in the adhesion at plaque–substrata interfaces, while DOPA-deficient proteins mfp-2, mfp-4, and mfp-6 show negligible adhesion to substrata [1,2,21,38,43, 47,61–67]. In this section, the recent progress on the interactions between mussel proteins and various substrata including mica, SiO_2, TiO_2, and polymers is reviewed.

3.3.2.1 Interaction between mussel foot proteins and mica

Mica is a hydrophilic aluminosilicate mineral, and as mentioned above it is atomically smooth in the basal plane (i.e., polysiloxane) with minor replacement of Si by Al (Fig. 3.3). The adhesion of mfps to mica is largely dependent on their DOPA content as measured by SFA. Mica-supported DOPA-deficient mfps (mfp-2, -4, -6) show negligible or weak adhesion to an opposing mica, while mica-supported DOPA-rich mfps (mfp-3, 5) show strong adhesion

[38, 57, 61]. No adhesion or weak adhesion of mfp-2, mfp-4 and mfp-6 in the asymmetric configuration shows that these proteins are not capable to bridge surfaces [38,60]. In contrast, mefp-3 shows strong adhesion and strong bridging ability to mica [62]. Mefp-5 achieves even stronger adhesion than mefp-3 and behaves like "superglue" [66]. Moderate adhesion was measured between *M. californianus* foot protein-1 (mcfp-1) film and mica, while oxidation of DOPA, high pH or high salt concentration could quickly reduce the adhesion [59,63]. Thus, the mfp-1 coating exposed to seawater (high ionic strength ~0.7 M, high pH ~8.2) could quickly lose its adhesive ability. The correlation between mfp adhesion and DOPA content was further explored by using mussel-inspired polypeptide analogues of mfp-3 with different DOPA contents of which the adhesion to mica increases with higher DOPA content [65]. In addition, DOPA oxidation by periodate treatment abolishes almost all of mfp-3 adhesion and more than 75% mfp-1 adhesion on mica [38,47,63].

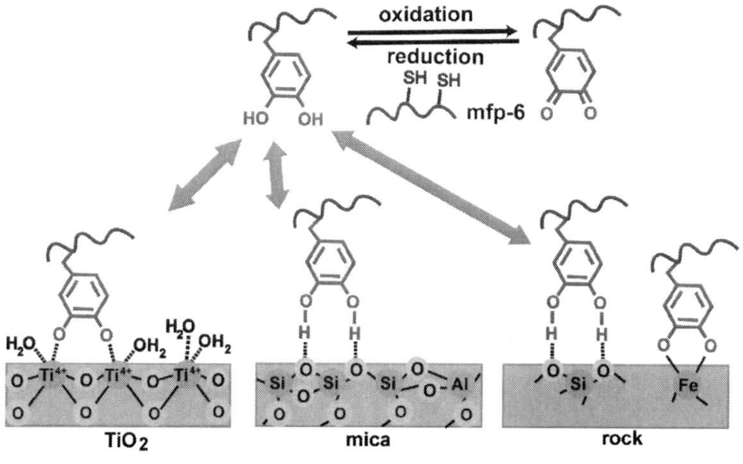

Figure 3.3 DOPA-mediated underwater adhesion of mussel proteins to various solid substrata (e.g., TiO_2, mica, rock) mainly by forming bidentate binuclear metal complexes and bidentate hydrogen bonding.

For the interaction mechanism, DOPA bidentate hydrogen bonding to mica (Fig. 3.3) has been proposed as the main contributor to the adhesion between mfps and mica due to the matching of the distance between the two hydroxyl groups in DOPA (0.29 nm) and the distance between adjacent O atoms in mica (0.28 nm) [59,65]. Metal complexation via oxidized Al groups interacting with phosphoester groups in mfp-5 and DOPA groups in all mfps may also play a role in their adhesion to mica [59]. Electrostatic attraction plays a minor role under the pH and salt concentrations studied [59].

3.3.2.2 Interaction between mussel foot proteins and silica

Silica or glass is relevant to the natural substrates (e.g., rock) in mussel habitats. Similar to the mica case, the adhesion of mfps to silica is highly dependent on DOPA. AFM force measurements showed that DOPA-rich proteins mfp-1 and mfp-5 adhere strongly to silica [10,11]. The mefp-1 adhesion to silica could be enhanced in the presence of multivalent cations (Mg^{2+}, Ca^{2+}, Fe^{3+}) [10]. Recombinant *M. galloprovincialis* foot protein 5 (mgfp-5) showed significant adhesion to glass in comparison to the adhesion of bovine serum albumin and commercial extracted mussel adhesive Cell-Tak [11]. SFA studies showed that mfp films (i.e., mcfp-1, mcfp-3 or mefp-5) adhere to silica surface, and the adhesion follows the order mfp-3 > mfp-5 ≈ mfp-1 at short contact time (i.e., minutes), which changes to mfp-3 > mfp-5 >> mfp-1 at relatively longer contact (i.e., 60 min) [59]. The proposed interaction mechanism of mfps on SiO_2 is very similar to the mica case: hydrogen bonding especially DOPA bidentate hydrogen bonding is the major contributor, while electrostatic interaction plays a minor role. The protein chain flexibility is also believed to play a role during its interaction to SiO_2 [59].

3.3.2.3 Interaction between mussel foot proteins and TiO_2/metal

The adhesion ability of mfps to TiO_2 surface is of great importance and has attracted much research interest due to the wide application of TiO_2 in bioengineering, such as dental and orthopedic implants. The adhesion of mfps to TiO_2 is also largely dependent on the DOPA chemistry, and the formation of bidentate binuclear complexes at the TiO_2–protein interface is proposed to be the main contributor for the adhesion (Fig. 3.3).

Force measurements of mfps or their mimics to TiO_2 have been directly measured by AFM and SFA. Lee et al. were the first to quantify the interaction energy between a single tethered DOPA and TiO_2 using AFM [12]. Strong adhesion was measured between DOPA and TiO_2, much stronger than that of DOPA to gold [12,68]. SFA studies showed that the wet adhesion of mussel-inspired polypeptide analogues of mfp-3 to TiO_2 increases with DOPA content [65]. Strong yet reversible wet adhesion was also measured between mcfp-1 and amorphous titania by SFA [43]. Oxidation of DOPA to dopaquinone reduces its adhesion to TiO_2 by 80% to 100%, indicating the dominant role of DOPA in adhesion. The presence of bidentate binuclear DOPA–Ti coordination bonds at the TiO_2–protein interface was identified by resonance Raman spectroscopy [43]. Similar bidentate binuclear complexes can also form between catechol and other metal oxides (Fe III, alumina) [12,43,67,69].

3.3.2.4 Interaction between mussel foot proteins and polymers

Understanding the interaction mechanisms of mfps to polymer substrata and biomaterials can shed light on the development of functional coatings, smart bioadhesives and drug delivery carriers. Both AFM and SFA have been applied to measure the interactions between mfps or their mimics and various polymer substrata.

The interaction of single DOPA residue and polyamine measured by AFM showed that DOPA can form covalent bonds with amine in basic solution [12]. It was reported that mussel-inspired four-armed poly(ethylene glycol) polymer functionalized with a single DOPA residue at the terminus of each arm PEG-$(DOPA)_4$ can significantly enhance the mucoadsorption of PEG, and strong adhesion was measured between DOPA and mucin [13]. Strong and reversible underwater adhesion was detected by AFM between tipless Si_3N_4 cantilever and gecko-inspired polymer pillars functionalized with a thin layer of mussel-inspired DOPA-rich polymer [14]. The interactions between thin mfp films (i.e., mcfp-1, mcfp-3 or mefp-5) and polymer substrata (i.e., polymethylmethacrylate (PMMA) and polystyrene (PS)) were directly probed by SFA. The SFA studies demonstrated that the mcfp-3 and mefp-5 show strong adhesion to PS surfaces, and moderate adhesion to PMMA surfaces, and mcfp-1 shows minor adhesion to both PS and PMMA surfaces [59]. The adhesion of mussel proteins to various polymer substrata is believed to be affected by several interaction mechanisms such as electrostatic interaction, hydrogen bonding, hydrophobic interactions, cation-π, π–π stacking [59]. Higher chain flexibility of mfps may enhance their adhesion because the local structure of the protein can better adapt to the specific surface chemistry [59].

It is evident from the above brief review that DOPA-substrate interactions significantly contribute to the mfp adhesion to various substrata, as illustrated in Fig. 3.3. For the substrata (i.e., rock) in mussel natural habitats, the dominant contribution to mussel adhesion is believed to be the formation of bidentate hydrogen bonding and iron-DOPA complexes at the rock–protein interfaces (Fig. 3.3). In seawater, DOPA itself can be easily oxidized to dopaquinone and quickly lose its adhesive ability. It was found that auto-oxidation of DOPA in mfp-3 occurs even at moderately acidic pH, indicating a certain degree of local redox and pH control present at the plaque–substrata interfaces for mussel adhesion in natural seawater [47]. It was proposed that an acidic reducing environment could be imposed based on the thiol-rich antioxidant protein mfp-6, under which DOPA could be restored at plaque–substratum interfaces (Fig. 3.3) [38].

3.4 Protein–Protein Interactions in Byssal Plaque

The mechanical properties of the byssus thread significantly rely on the molecular interactions (cohesion and adhesion) of the various proteins involved. The recent progress on the protein–protein interactions in the byssal plaque mainly includes: (1) cohesion mechanism of mfp-1s in the thin layer of cuticle and (2) cohesion and adhesion mechanism of various mfps within the plaque [7,21,64].

3.4.1 The Interaction of Mussel Foot Protein 1 (mfp-1) in the Cuticle

Mfp-1 is the single protein found in the mussel foot cuticle with a remarkable combination of high extensibility and high stiffness and hardness. Understanding the molecular interaction mechanism of mfp-1 can shed light on the exploration of bio-inspired wear and damage resistant medical coatings. Mfp-1 is composed of 75–85 repeats of identical decapeptides and contains significant amount of DOPA (10–15 mol%). Thus, mfp-1s can be easily cross-linked via chemical or enzymatic oxidation (e.g., using $NaIO_4$ and catechol oxidase) [9,70].

The molecular interaction mechanism accounting for the high extensibility and high stiffness of mussel cuticle was probed by SFA, which is mostly due to the strong and reversible metal ion bridging or chelation with DOPA in the cuticle. Fe^{3+} ions were found to be able to mediate strong and reversible bridging of non-adhesive symmetric thin mefp-1 films due to the formation of multiple bidentate complexation of iron by DOPA ligands [64]. However, the actual number of DOPA ligands bound can be affected by solution pH and the iron to DOPA ratio [64]. Metals such as iron, zinc, copper and manganese are normally concentrated in mussel plaque and tend to form DOPA–metal complexes. Recent Raman spectroscopy study showed that the byssal cuticle is a protein scaffold stabilized by DOPA–metal chelate complexes (mainly through iron) with unusual clustered distribution [7].

Substantial cohesion was also measured between DOPA-deficient foot proteins from green mussels *Perna viridis*, pvfp-1, with no metal complexation effects (i.e., Cu^{2+}, Fe^{3+}), suggesting the presence of other interaction mechanisms instead of metal chelation [61]. Different from mefp-1, pvfp-1 contains significant amount of modified amino acid C2-mannosyl-7-hydroxytryptophan (Man7OHTrp) instead of DOPA. The cation-π interactions with low desolvation penalty between π-electron-rich Man7OHTrp and lysine side chains were proposed to be the main contributor to the cohesion of pvfp-1 [61]. Cohesion of mcfp-1 films was also observed in the absence of metal ions, which was proposed to be mainly due to cation-π interactions between the positively charged lysine residues and π-electron-rich DOPA [63]. The addition of chemicals with aromatic groups (i.e., aspirin and 4-methylcatechol) can enhance the mcfp-1 cohesion [63]. The nanomechanics of cation-π interactions in aqueous solutions were directly probed by SFA using a model system with one surface coated with positively charged poly L-lysine (PLL) (primary amine, NH_3R^+) and the other surface coated with one of three homo-polymers with aromatic side groups: poly-L-tryptophan, poly-L-tyrosine, and PS [71]. The nanomechanical results further support the proposed cation-π interaction mechanism in mussel proteins (e.g., pvfp-1). The cation-π interactions in aqueous solutions were found to be dependent on the type and concentration of positively charged ions (e.g., tetraethylamonium, acetylcholine, Lys^+, NH_4^+, K^+, Li^+, and Na^+) and the type of aromatic side groups (e.g., indole, phenol, and benzene) [71]. It should be noted that cation-π interactions are highly orientation dependent. Perpendicular arrangement of cations on the π-electron-rich plane of the benzene ring is required to achieve efficient interactions.

3.4.1.1 Interactions between foot proteins in the plaque

The protein cohesion mechanism in the mussel plaque can provide important information for developing novel biomimetic composite materials. The interactions between two films of the same protein (mfp-2 vs. mfp-2, mfp-5 vs. mfp-5) or different proteins (mfp-2 vs. mfp-5, mfp-2 vs. mfp-3, mfp-2 vs. mfp-1, mfp-2 vs. mfp-4, mfp-3 vs. mfp-6) have been measured using SFA.

Mfp-2 is the most abundant plaque protein in the mussel foot plaque. Similar to the mefp-1 interaction, Fe^{3+} ions can induce strong and reversible bridging between two mefp-2 films due to the formation of multiple bidentate complexation of iron by DOPA ligands [21]. Ca^{2+} ions are also able to weakly bridge mfp-2 films through the interaction with some specific binding motif. The combination of strong DOPA-iron interaction and weak calcium binding provides a biological "safety net."

Strong cohesive interactions were measured between mfp-5 films and the cohesion strength decreased by 60% with increasing the salt concentration from 0.1 M to 0.35 M [60]. Mfp-5 contains positively charged residues and abundant aromatic content (~30 mol% DOPA). The adhesion could be attributed to cation-π interactions between positively charged lysine groups and aromatic DOPA groups.

In addition to the metal ion (e.g., Fe^{3+}, Ca^{2+})-mediated bridging, mfp-2 was found to bind strongly to mfp-5 at the plaque–substratum interface [21]. Similarly, Mfp-5 can interact with mfp-3. However, no obvious adhesion was detected between mfp-2 and mfp-3, or between mfp-2 and mfp-1. Ca^{2+}-mediated interaction between mfp-2 and mfp-4 was also detected. Recent SFA study suggested that mfp-6 may act as an efficient antioxidant and later a cross-linking partner for mfp-3, which is crucial for improving the cohesion among the plaque proteins [37,38,72]. Mfp-6 contains significant cysteinyl thiol content which can reduce dopaquinone back to

DOPA and recover the adhesion of mussel proteins (e.g., mfp-3) [38]. It should be noted that despite the above progress the protein–protein interaction mechanisms in the byssal plaque are still not completely understood [1,21].

The schematic in Fig. 3.4 summarizes the current understanding of intermolecular interactions among different mfps in the byssus. It is evident that DOPA plays a critical role in the cohesion and adhesion of mussel proteins via metal coordination (e.g., DOPA-Fe^{3+}), [7,21,64], cation-π interaction (e.g., DOPA-Lys^{+}), [61,63] and oxidative covalent cross-linking (e.g., DOPA-DOPA) [38,65,72–74]. More work is required to elucidate the full picture of intermolecular interaction mechanisms and quantify the structure–property relationships for the development of biomimetic materials.

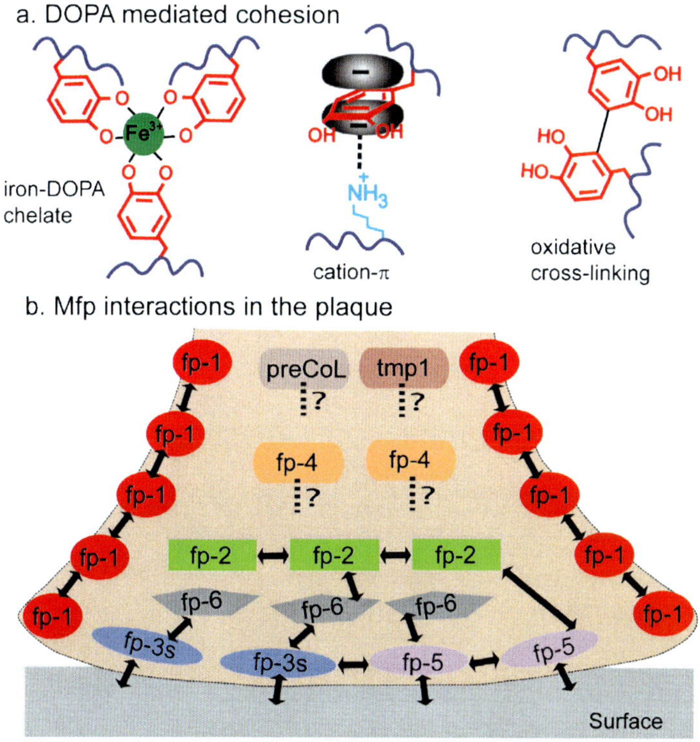

Figure 3.4 Schematic of protein–protein interactions in the byssal plaque: (a) DOPA-mediated cohesion mechanisms; (b) a proposed protein–protein interaction model.

3.5 Mussel-Inspired Materials

As discussed above, the strong and versatile wet adhesion capability of marine mussels is strongly dependent on the structures and molecular interactions of the various mussel proteins, particularly on the DOPA chemistry. The extensive experimental studies have shown several catechol interactions with DOPA involved in the wet adhesion process: (1) hydrogen bonding (with the ability to compete with water molecules), (2) metal–catechol coordination (e.g., Fe-$(DOPA)_3$ chelates yielding self-healing polymer networks), (3) formation of quinhydrone complexes largely stabilized by π-π interaction or cation-π interaction (e.g., DOPA-Lys^+), and (4) covalent bonding cross-linking by Schiff base and/or Michael addition reactions involving quinone groups of oxidized dopamine with nucleophilic groups like amine and thiol groups or direct free radical coupling (e.g., DOPA-DOPA mediating permanent polymer linkages). Inspired by these interaction mechanisms, bio-inspired materials with various engineering and biomedical applications have been developed. Some excellent reviews dedicated to the advances in biomimetic materials and their potential application can be found elsewhere. In this chapter, we mainly focus on the progress in two major areas of mussel-inspired materials: (1) adhesives/sealants and (2) functional coatings. Figure 3.5 summarizes some general approaches for developing mussel-inspired materials, as discussed further in the following sections.

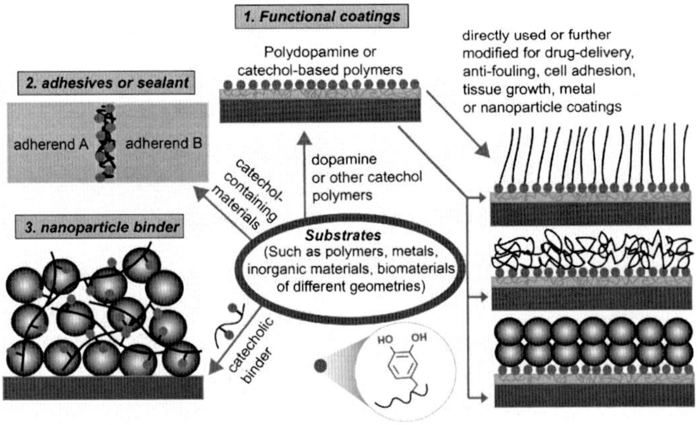

Figure 3.5 Schematic of mussel-inspired materials as adhesives or sealant, and functional coatings.

3.6 Adhesive/Sealants

The mussel's unique wet adhesion capability to virtually all kinds of substrates (viz., both inorganic and organic materials) has attracted much research interest in developing novel mussel-inspired adhesives or sealants. Previous studies on natural or synthetic DOPA-containing peptides show that DOPA plays a critical role in determining the wet adhesion performance of these biopolymers. DOPA is able to provide sufficient adhesion strength at the protein–substrate interface through various bonding mechanisms such as hydrogen bonding and catechol–metal chelating coordination, and DOPA is also easy to be oxidized by oxidants or enzymes to quinone, which can act as an effective cross-linking unit through Michael-type reactions with nucleophilic groups or by direct free radical aryl–aryl coupling [75–77]. The versatile DOPA chemistry and the low cytotoxicity and good biocompatibility of mfps provide a novel biomimetic approach for the development of adhesives and sealants with biomedical applications [78–81]. Since the pioneering work by Yamomoto et al. on the synthesis of water–soluble peptides containing *L*-DOPA via solution polymerization [82,83], significant efforts have been made to synthesize bio-mimetic peptides with similar amino acid sequence and DOPA content as mfps [76,77,84–86]. Synthetic adhesive hydrogels mimicking mfps have also been developed for potential biomedical applications, in which biocompatible and biodegradable polymers such as poly(ethylene oxide) (PEO), polycaprolactone, pluronic triblock copolymer are normally chosen as the hydrogel-forming materials [13,87–91]. In these studies, catechol groups were incorporated into different PEO-based linear and star-branched polymer frameworks. The solutions of these PEO polymers could readily form a hydrogel due to the formation of cross-linking units via the oxidation of catechol by oxidants or enzymes. The above mussel-inspired synthetic adhesives have been demonstrated to show strong adhesion to various tissues and overwhelm conventional fibrin glue and biomedical glue. As oxidants or enzymes are needed to induce a sol-gel transition, a two-component bio-inspired adhesive system is desirable which includes a mussel-inspired polymer and an oxidant or enzyme. For practical surgeon application, an injectable adhesive system with fast sol-gel transition is preferable. Recently complex coacervation or thermo-induced sol-gel transition has been introduced to the development of bio-inspired adhesive hydrogels. In the studies by Steward and co-workers [92,93], adhesive hydrogels were prepared through complex coacervation of two oppositely charged polymers. By tuning solution pH to a relatively basic pH, a gel-state coacervate could be formed mainly due to strong electrostatic interaction of the polymers. The addition of divalent cations (i.e., Ca^{2+} and Mg^{2+}) or exposing the DOPA-containing polymers to an oxidant, the cohesion strength of the adhesive gel can be enhanced [92,93]. Park and co-workers developed an injectable adhesive hydrogel by slightly cross-linking a thermo-responsive pluronic polymer with a biocompatible polymer hyaluronic acid or chitosan via Michael-type addition between thiol and catechol groups [94,95]. The adhesive hydrogel shows a fast and reversible temperature-dependent sol-gel transition and can be injected in vivo at room temperature via a syringe but swiftly becomes a robust gel in the body, which exhibits long-lasting stability and strong adhesion to biological tissues [94,95].

In addition to the covalent cross-linking mediated by the oxidation of catechol, strong yet reversible metal–catechol coordination (e.g., Fe^{3+}-DOPA) is another cross-linking strategy discovered in mussel proteins. Fe^{3+}-DOPA coordination (via mfp-1) was found to play an important role in the mechanical properties of the outer cuticle of byssal threads of marine mussels, which has several special features. The catechol-Fe^{3+} stoichiometry was found to be pH-dependent: monospecies dominate at pH < 5.6, bis at $5.6 < \text{pH} < 9.1$, and tris- at pH > 9.1 [96]. The bis- and tris-catechol-Fe^{3+} complexes are very stable with the equilibrium constants for the complex formation $\log K_S \approx 37–40$ [12,96]. The force required to break a metal–catechol bond (e.g., a typical value of ~0.8 nN) is close to that to rupture a covalent bond (e.g., a typical value of ~2 nN) under identical loading conditions based on single molecule force measurements [12,96]. The strong and reversible metal–catechol interaction mechanism has inspired the development of novel self-healing materials and adhesives for biomedical applications. Both catechol-mediated coordination cross-linking and pH-mediated cross-linking have been introduced to modulate the self-healing properties of the materials. Messersmith and co-workers developed a method to modulate catechol-Fe^{3+} cross-linking via controlling pH, and the resultant hydrogel shows self-healing properties with elastic moduli close to covalently cross-linked counterpart [96,97]. Very recently, a self-healing mussel-inspired multi-pH-responsive hydrogel has been developed using DOPA-functionalized polyallylamine following a similar strategy as mentioned above [98]. In this study, the degree of polymer cross-linking can be controlled through the pH-dependent DOPA-iron coordination interaction, and the hydrogels show maximum mechanical strength at a pH close to the PI value of the polymer (viz. pK_a value of the amine side groups of polyallylamine) [98]. Therefore, in principle, the development of self-healing hydrogels with maximum mechanical strength at desired pH can be achieved by following a mussel-inspired approach and modulating the pK_a values of the polymer. Photoresponsive mussel-inspired hydrogel has also been developed by del Campo and co-workers by introducing 5-nitro-dopamine moieties instead of dopamine into the hydrogel, which show both self-healing and light-triggered debonding/degradation properties [99]. The synthesis process of polypeptide structure in biomimetic polymers is normally expensive and time-consuming. Other structures have been also adopted in the biomimetic adhesives, such as polystyrene, polyacrylate and polyacrylamide prepared by free radical polymerization and natural polysaccharides (e.g., chitosan). For example, Wilker and co-workers synthesized catechol-containing polystyrene, and investigated the impact of polymer composition (e.g., DOPA, cationic groups) and substrate on the adhesion after curing [100–103]. Takahara and co-workers developed a new type of mussel-mimic hydrogel adhesives based on acrylamide-type copolymers containing catechol, amino, and hydroxyl groups, and investigated the lap shear strength of the adhesives [104]. Stepuk et al. developed a co-polymer of methyl methacrylate and mussel-inspired dopamine methacrylamide, and an adhesion strength of up to 20 MPa was obtained by using the adhesives to bond polymers to Ti [105].

The unique hierarchical structures of gecko toes allow the gecko to possess fast reversible attachment and detachment during its movement, which has also attracted much research interest as a smart dry adhesive. Recently, gecko-inspired polymer pillars functionalized with a thin layer of mussel-inspired DOPA-rich polymers have been fabricated and show good adhesion performance in both dry and wet conditions [14,106].

3.7 Functional Coating

Dopamine and its analogs can self-polymerize readily and have strong coating capability to a wide range of organic and inorganic substrates, which provides a robust and general method to prepare functional coatings [3]. The adherent poly(dopamine) (PDA) film coatings are composed of a mixture of cross-linked products bearing mostly catechol and quinone groups, which can serve as a versatile platform for secondary surface-mediated reactions. One of the important applications of the above mussel-inspired coating strategy is for the development of antifouling coatings. Messersmith and co-workers developed bifunctional polymers by coupling mussel-mimetic anchors and antifouling polymers such as PEO [67,68,107,108] or *N*-methoxyethylglycine (peptoid) oligomer [109]. The mussel-mimetic anchors can be either short polypeptide chains consisting of DOPA and lysine residues, or anchoring moieties containing DOPA residue. The treated substrates (e.g., TiO_2 and Ti) using these functional polymers show excellent antifouling properties of resisting the attachment of proteins and cells. Surface-initiated atom transfer radical polymerization (ATRP) by using a catecholic initiator has also been applied in development of antifouling surfaces. A Ti substrate was first functionalized using a DOPA-based initiator, and subsequently reacted with oligo-poly(ethelene oxide) methacrylate to form a polymer brush antifouling coating. Photo-mask lithography was further adapted to pattern the treated Ti surfaces with alternative antifouling and fouling areas [110]. Following a similar approach, Textor and co-workers reported the synthesis of an antifouling coating by using dendritic PEO polymers conjugated with catechol groups [111]. It is noted that the DOPA can be easily oxidized which impacts the long-term stability of the coatings. Gademann and co-workers developed a series of synthetic analogs of the catecholic chromophore moiety of siderophore anachelinone, which were conjugated to PEO [112]. Due to the presence of a quaternized amino group, such DOPA analog anchor shows better anti-oxidation ability than that of DOPA-based moieties; meanwhile, the positive charge on the anchor further enhance its affinity to negatively charged substrates, resulting in a better and more stable antifouling coating [112]. Multiple DOPA or DOPA analogs have been grafted to polymer backbones to strengthen their anchoring ability to substrates. Catecholic moieties

(3,4-dihydroxyphenylacetic acid, DHPAA) and PEO were grafted to polylysine backbone, and the polymer was coated on titanium oxide surfaces, which exhibits excellent antifouling resistance to nonspecific adsorption of fibrinogen protein in high ionic salt solutions and show long-term stability (>100 days) [113]. Catechol-grafted poly(ethylene) glycol (PEG-g-catechol) for the preparation of nonfouling coating on versatile substrates including Au, Ti, Si, poly(carbonate) and poly(tetrafluoro ethylene) (PTFE), which were very resistant to proteins and cells [114].

Magnetic nanoparticles (MNP) have attracted much attention due to their broad biomedical applications ranging from magnetic resonance imaging (MRI), hyperthermia, and drug delivery to biosensing and biochemical separation. One prerequisite for the medical applications of MNP is that the MNP should be stable in aqueous solution and homogenously dispersed. Due to the strong coordination ability to the constituted metal ions of MNP such as Fe^{3+}, Mn^{2+}, and Mn^{3+} [7,42,115–117], catechol derivatives have been widely used as robust surface anchors to immobilize functional molecules/polymers on the MNP [118,119]. Polymer ligands possessing multiple mussel-inspired anchoring groups have been applied to enhance the affinity of these ligands to nanocrystal surfaces, which are stable in complex biological conditions [120, 121]. Nitrocatechols and chlorinated catechols have also been used as anchors for polymer ligands to improve their oxidation resistance [122]. The above treated nanoparticles (with the mussel-mimetic coatings) typically show high stability even at harsh environment (e.g., elevated temperature, high ionic strength) for the various applications.

The PDA coatings have also been applied for the film deposition of gold or silver nanoparticles [74,123,124] and to reduce graphene oxide [125–127] without the use of an exogenous reducing agent to fabricate hybrid films on a wide range of substrates. PAD-based PDA capsule and carbon capsule were prepared as a potential carrier for drug delivery and catalyst template, respectively [128].

3.8 Conclusions

Marine mussels have attracted increased research interest in the fields of adhesion science and materials science due to their excellent adhesion ability to a wide variety of substrates underwater.

Significant progress has been made toward the understanding of the associated wet adhesion mechanism. Nanomechanical measurements through SFA and AFM have revealed the molecular interaction mechanisms of mussel adhesive proteins in complex aqueous environment. The wet adhesion mechanisms discovered have inspired the development of biomimetic materials and devices with various important applications such as biomedical and dental adhesives, self-healing materials, stimuli-responsive materials, anti-fouling coatings and biosensors. Further efforts are needed to better understand the catechol-based materials chemistry, and optimize the molecular structures and performances of bio-inspired materials for the different applications.

Acknowledgment

This work was supported by Natural Sciences and Engineering Research Council of Canada (NSERC) and Canada Foundation for Innovation (CFI).

References

1. Lee, B. P., Messersmith, P. B., Israelachvili, J. N., and Waite, J. H. (2011). Mussel-inspired adhesives and coatings, *Ann. Rev. Mater. Res.*, **41**, 99–132.
2. Holten-Andersen, N., and Waite, J. H. (2008). Mussel-designed protective coatings for compliant substrates, *J. Dent. Res.*, **87**, 701–709.
3. Lee, H., Dellatore, S. M., Miller, W. M., and Messersmith, P. B. (2007). Mussel-inspired surface chemistry for multifunctional coatings, *Science*, **318**, 426–430.
4. Kamino, K. (2008). Underwater adhesive of marine organisms as the vital link between biological science and material science, *Mar. Biotechnol.*, **10**, 111–121.
5. Waite, J. H., Andersen, N. H., Jewhurst, S., and Sun, C. J. (2005). Mussel adhesion: Finding the tricks worth mimicking, *J. Adhesion.*, **81**, 297–317.
6. Brown, C. H. (1952). Some structural proteins of Mytilus-edulis, *Q. J. Microsc. Sci.*, **93**, 487–502.
7. Harrington, M. J., Masic, A., Holten-Andersen, N., Waite, J. H., and Fratzl, P. (2010). Iron-clad fibers: a metal-based biological strategy for hard flexible coatings, *Science*, **328**, 216–220.

8. Fant, C., Hedlund, J., Hook, F., Berglin, M., Fridell, E., and Elwing, H. (2010). Investigation of adsorption and cross-linking of a mussel adhesive protein using attenuated total internal reflection fourier transform infrared spectroscopy (ATR-FTIR), *J. Adhesion.*, **86**, 25–38.
9. Hook, F., Kasemo, B., Nylander, T., Fant, C., Sott, K., and Elwing, H. (2001). Variations in coupled water, viscoelastic properties, and film thickness of a Mefp-1 protein film during adsorption and cross-linking: A quartz crystal microbalance with dissipation monitoring, ellipsometry, and surface plasmon resonance study, *Anal. Chem.*, **73**, 5796–5804.
10. Frank, B. P., and Belfort, G. (2002). Adhesion of Mytilus edulis foot protein 1 on silica: Ionic effects on bifouling, *Biotechnol. Prog.*, **18**, 580–586.
11. Hwang, D. S., Yoo, H. J., Jun, J. H., Moon, W. K., and Cha, H. J. (2004). Expression of functional recombinant mussel adhesive protein Mgfp-5 in Escherichia coli, *Appl. Environ. Microb.*, **70**, 3352–3359.
12. Lee, H., Scherer, N. F., and Messersmith, P. B. (2006). Single-molecule mechanics of mussel adhesion, *P. Natl. Acad. Sci. U. S. A.*, **103**, 12999–13003.
13. Catron, N. D., Lee, H., and Messersmith, P. B. (2006). Enhancement of poly(ethylene glycol) mucoadsorption by biomimetic end group functionalization, *Biointerphases*, **1**, 134–141.
14. Lee, H., Lee, B. P., and Messersmith, P. B. (2007). A reversible wet/dry adhesive inspired by mussels and geckos, *Nature*, **448**, 338–341.
15. Kavouras, J. H., and Maki, J. S. (2003). Effects of biofilms on zebra mussel postveliger attachment to artificial surfaces, *Invertebr. Biol.*, **122**, 138–151.
16. Marsden, J. E., and Lansky, D. M. (2000). Substrate selection by settling zebra mussels, Dreissena polymorpha, relative to material, texture, orientation, and sunlight, *Can. J. Zool.*, **78**, 787–793.
17. Kavouras, J. H., and Maki, J. S. (2004). Inhibition of the reattachment of young adult zebra mussels by single-species biofilms and associated exopolymers, *J. Appl. Microbiol.*, **97**, 1236–1246.
18. Waite, J. H. (1987). Natures underwater adhesive specialist, *Int. J. Adhes. Adhes.*, **7**, 9–14.
19. Waite, J. H. (1992). The formation of mussel byssus: Anatomy of a natural manufacturing process., *Results Probl. Cell Differ.*, **19**, 27–54.
20. Qin, X. X., and Waite, J. H. (1995). Exotic collagen gradients in the byssus of the mussel Mytilus-edulis, *J. Exp. Biol.*, **198**, 633–644.

21. Hwang, D. S., Zeng, H. B., Masic, A., Harrington, M. J., Israelachvili, J. N., and Waite, J. H. (2010). Protein- and metal-dependent interactions of a mrominent protein in mussel adhesive plaques, *J. Biol. Chem.*, **285**, 25850–25858.
22. Holten-Andersen, N., Mates, T. E., Toprak, M. S., Stucky, G. D., Zok, F. W., and Waite, J. H. (2009). Metals and the integrity of a biological coating: the cuticle of mussel byssus, *Langmuir*, **25**, 3323–3326.
23. Filpula, D. R., Lee, S. M., Link, R. P., Strausberg, S. L., and Strausberg, R. L. (1990). Structural and functional repetition in a marine mussel adhesive protein, *Biotechnol. Prog.*, **6**, 171–177.
24. Waite,J.H.(1983).Evidenceforarepeating3,4-Dihydroxyphenylalanine-containing and hydroxyproline-containing decapeptide in the adhesive protein of the mussel, Mytilus-edulis-L, *J. Biol. Chem.*, **258**, 2911–2915.
25. Taylor, S. W., Waite, J. H., Ross, M. M., Shabanowitz, J., and Hunt, D. F. (1994). Trans-2,3-*cis*-3,4-dihydroxyproline, a new naturally-occurring amino-acid, Is the 6th residue in the tandemly repeated consensus decapeptides of an adhesive protein from Mytilus-edulis, *J. Am. Chem. Soc.*, **116**, 10803–10804.
26. Rzepecki, L. M., Hansen, K. M., and Waite, J. H. (1992). Characterization of a cystine-rich polyphenolic protein family from the blue mussel mytilus-edulis-L, *Biol. Bull-US.*, **183**, 123–137.
27. Waite, J. H. (1986). Mussel glue from Mytilus-Californianus conrad—a comparative-study, *J. Comp. Physiol. B, Biochem. Syst. Environ. Physiol.*, **156**, 491–496.
28. Inoue, K., Takeuchi, Y., Miki, D., and Odo, S. (1995). Mussel adhesive plaque protein gene is a novel member of epidermal growth factor-like gene family, *J. Biol. Chem.*, **270**, 6698–6701.
29. Haemers, S., van der Leeden, M. C., and Frens, G. (2005). Coil dimensions of the mussel adhesive protein Mefp-1, *Biomaterials*, **26**, 1231–1236.
30. Papov, V. V., Diamond, T. V., Biemann, K., and Waite, J. H. (1995). Hydroxyarginine-containing polyphenolic proteins in the adhesive plaques of the marine mussel Mytilus-edulis, *J. Biol. Chem.*, **270**, 20183–20192.
31. Warner, S. C., and Waite, J. H. (1999). Expression of multiple forms of an adhesive plaque protein in an individual mussel, Mytilus edulis, *Mar. Biol.*, **134**, 729–734.
32. Even, M. A., Wang, J., and Chen, Z. (2008). Structural information of mussel adhesive protein Mefp-3 acquired at various polymer/Mefp-3 solution interfaces, *Langmuir*, **24**, 5795–5801.

33. Burzio, L. A., and Waite, J. H. (2000). Cross-linking in adhesive quinoproteins: Studies with model decapeptides, *Biochemistry-US.*, **39**, 11147–11153.
34. Zhao, H., and Waite, J. H. (2006). Proteins in load-bearing junctions: The histidine-rich metal-binding protein of mussel byssus, *Biochemistry-US.*, **45**, 14223–14231.
35. Waite, J. H., and Qin, X. X. (2001). Polyphosphoprotein from the adhesive pads of Mytilus edulis, *Biochemistry-US.*, **40**, 2887–2893.
36. Cha, H. J., Hwang, D. S., and Lim, S. (2008). Development of bioadhesives from marine mussels, *Biotech. J.*, **3**, 631–638.
37. Zhao, H., and Waite, J. H. (2006). Linking adhesive and structural proteins in the attachment plaque of Mytilus californianus, *J. Biol. Chem.*, **281**, 26150–26158.
38. Yu, J., Wei, W., Danner, E., Ashley, R. K., Israelachvili, J. N., and Waite, J. H. (2011). Mussel protein adhesion depends on interprotein thiol-mediated redox modulation, *Nat. Chem. Biol.*, **7**, 588–590.
39. Silverman, H. G., and Roberto, F. F. (2007). Understanding marine mussel adhesion, *Mar. Biotechnol.*, **9**, 661–681.
40. Bandara, N., Zeng, H., and Wu, J. (2012). Marine mussel adhesion: Biochemistry, mechanisms, and biomimetics, *J. Adhes. Sci. Technol.*, 1–24.
41. Fant, C., Sott, K., Elwing, H., and Hook, F. (2000). Adsorption behavior and enzymatically or chemically induced cross-linking of a mussel adhesive protein, *Biofouling*, **16**, 119–132.
42. Taylor, S. W., Chase, D. B., Emptage, M. H., Nelson, M. J., and Waite, J. H. (1996). Ferric ion complexes of a DOPA-containing adhesive protein from Mytilus edulis, *Inorg. Chem.*, **35**, 7572–7577.
43. Hwang, D. S., Harrington, M. J., Lu, Q. Y., Masic, A., Zeng, H. B., and Waite, J. H. (2012). Mussel foot protein-1 (mcfp-1) interaction with titania surfaces, *J. Mater. Chem.*, **22**, 15530–15533.
44. Suci, P. A., and Geesey, G. G. (1995). Investigation of alginate binding to germanium and polystyrene substrata conditioned with mussel adhesive protein, *J. Colloid Interf. Sci.*, **172**, 347–357.
45. Suci, P. A., and Geesey, G. G. (2000). Influence of sodium periodate and tyrosinase on binding of alginate to adlayers of Mytilus edulis foot protein 1, *J. Colloid Interf. Sci.*, **230**, 340–348.
46. Suci, P. A., and Geesey, G. G. (2001). Use of attenuated total internal reflection Fourier transform infrared spectroscopy to investigate interactions between Mytilus edulis foot proteins at a surface, *Langmuir*, **17**, 2538–2540.

47. Yu, J., Wei, W., Danner, E., Israelachvili, J. N., and Waite, J. H. (2011). Effects of interfacial redox in mussel adhesive protein films on mica, *Adv. Mater.*, **23**, 2362–2366.
48. Hedlund, J., Andersson, M., Fant, C., Bitton, R., Bianco-Peled, H., Elwing, H., and Berglin, M. (2009). Change of colloidal and surface properties of Mytilus edulis foot protein 1 in the presence of an oxidation ($NaIO_4$) or a complex-binding (Cu^{2+}) agent, *Biomacromolecules*, **10**, 845–849.
49. Sever, M. J., Weisser, J. T., Monahan, J., Srinivasan, S., and Wilker, J. J. (2004). Metal-mediated cross-linking in the generation of a marine-mussel adhesive, *Angew. Chem. Int. Edit.*, **43**, 448–450.
50. Israelachvili, J. N., and Mcguiggan, P. M. (1990). Adhesion and short-range forces between surfaces. 1. New apparatus for surface force measurements, *J. Mater. Res.*, **5**, 2223–2231.
51. Israelachvili, J., Min, Y., Akbulut, M., Alig, A., Carver, G., Greene, W., Kristiansen, K., Meyer, E., Pesika, N., Rosenberg, K., and Zeng, H. (2010). Recent advances in the surface forces apparatus (SFA) technique, *Rep. Prog. Phys.*, **73**, 1–16.
52. Tabor, D., and Winterton, R. H. S. (1969). Direct measurement of normal and retarded Van Der Waals forces, *Proc. R. Soc. London, Ser. A.*, **312**, 435–450.
53. Israelachvili, J. N., and Tabor, D. (1972). Measurement of Van der Waals dispersion forces in range 1.5 to 130 nm, *Proc. R. Soc. London, Ser. A.*, **331**, 19–38.
54. Israelachvili, J. N., and Adams, G. E. (1976). Direct measurement of long-range forces between 2 mica surfaces in aqueous KNO_3 solutions, *Nature*, **262**, 773–776.
55. Israelachvili, J. N., and Adams, G. E. (1978). Measurement of forces between 2 Mica surfaces in aqueous-electrolyte solutions in range 0–100 nm, *J. Chem. Soc. Faraday Trans. I.*, **74**, 975–1001.
56. Zeng, H., Tian, Y., Anderson, T. H., Tirrell, M., and Israelachvili, J. N. (2008). New SFA techniques for studying surface forces and thin film patterns induced by electric fields, *Langmuir*, **24**, 1173–1182.
57. Israelachvili, J. N. (1973). Thin-film studies using multiple-beam interferometry, *J. Colloid Interf. Sci.*, **44**, 259–272.
58. Israelachvili, J. N., (2011). *Intermolecular and Surface Forces*, 3rd. ed. (Elsevier Academic Press, USA) pp. 1–674.
59. Lu, Q. Y., Danner, E., Waite, J. H., Israelachvili, J. N., Zeng, H. B., and Hwang, D. S. (2013). Adhesion of mussel foot proteins to different substrate surfaces, *J. R. Soc. Interface*, **10**, 20120759.

60. Hwang, D. S., Wei, W., Rodriguez-Martinez, N. R., Danner, E., and Waite, J. H., (2013), A microcosm of wet adhesion: dissecting protein interactions in mussel attachment plaques, in *Polymer Adhesion, Friction, and Lubrication* (Zeng, H., ed.), John Wiley & Sons,USA, Chapter 8, pp. 319–349.

61. Hwang, D. S., Zeng, H. B., Lu, Q. Y., Israelachvili, J., and Waite, J. H. (2012). Adhesion mechanism in a DOPA-deficient foot protein from green mussels, *Soft Matter.*, **8**, 5640–5648.

62. Lin, Q., Gourdon, D., Sun, C. J., Holten-Andersen, N., Anderson, T. H., Waite, J. H., and Israelachvili, J. N. (2007). Adhesion mechanisms of the mussel foot proteins mfp-1 and mfp-3, *P. Natl. Acad. Sci. U. S. A.*, **104**, 3782–3786.

63. Lu, Q. Y., Hwang, D. S., Liu, Y., and Zeng, H. B. (2012). Molecular interactions of mussel protective coating protein, mcfp-1, from Mytilus californianus, *Biomaterials*, **33**, 1903–1911.

64. Zeng, H. B., Hwang, D. S., Israelachvili, J. N., and Waite, J. H. (2010). Strong reversible Fe^{3+}-mediated bridging between dopa-containing protein films in water, *P. Natl. Acad. Sci. U. S. A.*, **107**, 12850–12853.

65. Anderson, T. H., Yu, J., Estrada, A., Hammer, M. U., Waite, J. H., and Israelachvili, J. N. (2010). The contribution of DOPA to substrate-peptide adhesion and internal cohesion of mussel-inspired synthetic peptide films, *Adv. Funct. Mater.*, **20**, 4196–4205.

66. Danner, E. W., Kan, Y. J., Hammer, M. U., Israelachvili, J. N., and Waite, J. H. (2012). Adhesion of mussel foot protein Mefp-5 to mica: An underwater superglue, *Biochemistry-US.*, **51**, 6511–6518.

67. Dalsin, J. L., Lin, L. J., Tosatti, S., Voros, J., Textor, M., and Messersmith, P. B. (2005). Protein resistance of titanium oxide surfaces modified by biologically inspired mPEG-DOPA, *Langmuir*, **21**, 640–646.

68. Dalsin, J. L., Hu, B. H., Lee, B. P., and Messersmith, P. B. (2003). Mussel adhesive protein mimetic polymers for the preparation of nonfouling surfaces, *J. Am. Chem. Soc.*, **125**, 4253–4258.

69. Mcbride, M. B., and Wesselink, L. G. (1988). Chemisorption of catechol on gibbsite, boehmite, and noncrystalline alumina surfaces, *Environ. Sci. Technol.*, **22**, 703–708.

70. Waite, J. H., and Tanzer, M. L. (1981). Polyphenolic substance of Mytilus-edulis - novel adhesive containing L-Dopa and hydroxyproline, *Science*, **212**, 1038–1040.

71. Lu, Q. Y., Oh, D. X., Lee, Y., Jho, Y., Hwang, D. S., and Zeng, H. B. (2013). Nanomechanics of cation interactions in aqueous solution, *Angew. Chem. Int. Edit.*, **52**, 3944–3948.

72. McDowell, L. M., Burzio, L. A., Waite, J. H., and Schaefer, J. (1999). Rotational echo double resonance detection of cross-links formed in mussel byssus under high-flow stress, *J. Biol. Chem.*, **274**, 20293–20295.

73. Stewart, R. J., Ransom, T. C., and Hlady, V. (2011). Natural underwater adhesives, *J. Polym. Sci. Polym. Phys.*, **49**, 757–771.

74. Wilker, J. J. (2010). Marine bioinorganic materials: Mussels pumping iron, *Curr. Opin. Chem. Biol.*, **14**, 276–283.

75. Waite, J. H. (2002). Adhesion a la moule, *Integ. Comp. Biol.*, **42**, 1172–1180.

76. Yu, M. E., and Deming, T. J. (1998). Synthetic polypeptide mimics of marine adhesives, *Macromolecules*, **31**, 4739–4745.

77. Yu, M. E., Hwang, J. Y., and Deming, T. J. (1999). Role of L-3,4-dihydroxyphenylalanine in mussel adhesive proteins, *J. Am. Chem. Soc.*, **121**, 5825–5826.

78. Schnurrer, J., and Lehr, C. M. (1996). Mucoadhesive properties of the mussel adhesive protein, *Int. J. Pharm.*, **141**, 251–256.

79. Chivers, R. A., and Wolowacz, R. G. (1997). The strength of adhesive-bonded tissue joints, *Int. J. Adhes. Adhes.*, **17**, 127–132.

80. Ninan, L., Monahan, J., Stroshine, R. L., Wilker, J. J., and Shi, R. Y. (2003). Adhesive strength of marine mussel extracts on porcine skin, *Biomaterials*, **24**, 4091–4099.

81. Ninan, L., Stroshine, R. L., Wilker, J. J., and Shi, R. (2007). Adhesive strength and curing rate of marine mussel protein extracts on porcine small intestinal submucosa, *Acta Biomater.*, **3**, 687–694.

82. Yamamoto, H., and Hayakawa, T. (1982). Synthesis of sequential polypeptides containing L-Beta-3,4-Dihydroxyphenyl-Alpha-Alanine (Dopa) and L-Lysine, *Biopolymers*, **21**, 1137–1151.

83. Yamamoto, H. (1987). Synthesis and adhesive studies of marine polypeptides, *J. Chem. Soc. Perkin. Trans.*, **1**, 613–618.

84. Sever, M. J., and Wilker, J. J. (2001). Synthesis of peptides containing DOPA (3,4-dihydroxyphenylalanine), *Tetrahedron*, **57**, 6139–6146.

85. Wang, J., Liu, C. S., Lu, X., and Yin, M. (2007). Co-polypeptides of 3,4-dihydroxyphenylalanine and L-lysine to mimic marine adhesive protein, *Biomaterials*, **28**, 3456–3468.

86. Deming, T. J. (2007). Synthetic polypeptides for biomedical applications, *Prog. Polym. Sci.*, **32**, 858–875.

87. Huang, K., Lee, B. P., Ingram, D. R., and Messersmith, P. B. (2002). Synthesis and characterization of self-assembling block copolymers containing bioadhesive end groups, *Biomacromolecules*, **3**, 397–406.

88. Lee, B. P., Dalsin, J. L., and Messersmith, P. B. (2002). Synthesis and gelation of DOPA-Modified poly(ethylene glycol) hydrogels, *Biomacromolecules*, **3**, 1038–1047.

89. Hu, B. H., and Messersmith, P. B. (2003). Rational design of transglutaminase substrate peptides for rapid enzymatic formation of hydrogels, *J. Am. Chem. Soc.*, **125**, 14298–14299.

90. Lee, B. P., Chao, C. Y., Nunalee, F. N., Motan, E., Shull, K. R., and Messersmith, P. B. (2006). Rapid gel formation and adhesion in photocurable and biodegradable block copolymers with high DOPA content, *Macromolecules*, **39**, 1740–1748.

91. Guvendiren, M., Messersmith, P. B., and Shull, K. R. (2008). Self-assembly and adhesion of DOPA-modified methacrylic triblock hydrogels, *Biomacromolecules*, **9**, 122–128.

92. Shao, H., and Stewart, R. J. (2010). Biomimetic underwater adhesives with environmentally triggered setting mechanisms, *Adv. Mater.*, **22**, 729–733.

93. Shao, H., Bachus, K. N., and Stewart, R. J. (2009). A water-borne adhesive modeled after the sandcastle glue of P-californica, *Macromol. Biosci.*, **9**, 464–471.

94. Lee, Y., Chung, H. J., Yeo, S., Ahn, C. H., Lee, H., Messersmith, P. B., and Park, T. G. (2010). Thermo-sensitive, injectable, and tissue adhesive sol-gel transition hyaluronic acid/pluronic composite hydrogels prepared from bio-inspired catechol-thiol reaction, *Soft Matter.*, **6**, 977–983.

95. Ryu, J. H., Lee, Y., Kong, W. H., Kim, T. G., Park, T. G., and Lee, H. (2011). Catechol-functionalized chitosan/pluronic hydrogels for tissue adhesives and hemostatic materials, *Biomacromolecules*, **12**, 2653–2659.

96. Holten-Andersen, N., Harrington, M. J., Birkedal, H., Lee, B. P., Messersmith, P. B., Lee, K. Y. C., and Waite, J. H. (2011). pH-induced metal-ligand cross-links inspired by mussel yield self-healing polymer networks with near-covalent elastic moduli, *P. Natl. Acad. Sci. U. S. A.*, **108**, 2651–2655.

97. Ceylan, H., Urel, M., Erkal, T. S., Tekinay, A. B., Dana, A., and Guler, M. O. (2013). Mussel inspired dynamic cross-linking of self-healing peptide nanofiber network, *Adv. Funct. Mater.*, **23**, 2081–2090.

98. Krogsgaard, M., Behrens, M. A., Pedersen, J. S., and Birkedal, H. (2013). Self-healing mussel-inspired multi-pH-responsive hydrogels, *Biomacromolecules*, **14**, 297–301.

99. Shafiq, Z., Cui, J. X., Pastor-Perez, L., San Miguel, V., Gropeanu, R. A., Serrano, C., and del Campo, A. (2012). Bioinspired underwater bonding and debonding on demand, *Angew. Chem. Int. Edit.*, **51**, 4332–4335.

100. Westwood, G., Horton, T. N., and Wilker, J. J. (2007). Simplified polymer mimics of cross-linking adhesive proteins, *Macromolecules*, **40**, 3960–3964.

101. White, J. D., and Wilker, J. J. (2011). Underwater bonding with charged polymer mimics of marine mussel adhesive proteins, *Macromolecules*, **44**, 5085–5088.

102. Jenkins, C. L., Meredith, H. J., and Wilker, J. J. (2013). Molecular weight effects upon the adhesive bonding of a mussel mimetic polymer, *ACS Appl. Mater. Interfaces*, **5**, 5091–5096.

103. Matos-Perez, C. R., White, J. D., and Wilker, J. J. (2012). Polymer composition and substrate influences on the adhesive bonding of a biomimetic, cross-linking polymer, *J. Am. Chem. Soc.*, **134**, 9498–9505.

104. Nishida, J., Kobayashi, M., and Takahara, A. (2013). Gelation and adhesion behavior of mussel adhesive protein mimetic polymer, *J. Polym. Sci. Polym. Chem.*, **51**, 1058–1065.

105. Stepuk, A., Halter, J. G., Schaetz, A., Grass, R. N., and Stark, W. J. (2012). Mussel-inspired load bearing metal-polymer glues, *Chem. Commun.*, **48**, 6238–6240.

106. Glass, P., Chung, H. Y., Washburn, N. R., and Sitti, M. (2010). Enhanced wet adhesion and shear of elastomeric micro-fiber arrays with mushroom tip geometry and a photopolymerized p(DMA-co-MEA) tip coating, *Langmuir*, **26**, 17357–17362.

107. Fan, X. W., Lin, L. J., and Messersmith, P. B. (2006). Cell fouling resistance of polymer brushes grafted from Ti substrates by surface-initiated polymerization: Effect of ethylene glycol side chain length, *Biomacromolecules*, **7**, 2443–2448.

108. Gunawan, R. C., King, J. A., Lee, B. P., Messersmith, P. B., and Miller, W. M. (2007). Surface presentation of bioactive Ligands in a nonadhesive background using DOPA-Tethered biotinylated poly(ethylene glycol), *Langmuir*, **23**, 10635–10643.

109. Statz, A. R., Meagher, R. J., Barron, A. E., and Messersmith, P. B. (2005). New peptidomimetic polymers for antifouling surfaces, *J. Am. Chem. Soc.*, **127**, 7972–7973.

110. Fan, X. W., Lin, L. J., Dalsin, J. L., and Messersmith, P. B. (2005). Biomimetic anchor for surface-initiated polymerization from metal substrates, *J. Am. Chem. Soc.*, **127**, 15843–15847.

111. Gillich, T., Benetti, E. M., Rakhmatullina, E., Konradi, R., Li, W., Zhang, A., Schluter, A. D., and Textor, M. (2011). Self-assembly of focal point Oligo-catechol ethylene glycol dendrons on Titanium oxide surfaces: Adsorption kinetics, surface characterization, and nonfouling Properties, *J. Am. Chem. Soc.*, **133**, 10940–10950.

112. Wach, J. Y., Bonazzi, S., and Gademann, K. (2008). Antimicrobial surfaces through natural product hybrids, *Angew. Chem. Int. Edit.*, **47**, 7123–7126.

113. Saxer, S., Portmann, C., Tosatti, S., Gademann, K., Zurcher, S., and Textor, M. (2010). Surface assembly of Catechol-functionalized poly(L-lysine)-graft-poly(ethylene glycol) copolymer on Titanium exploiting combined electrostatically driven self-organization and blomimetic strong adhesion, *Macromolecules*, **43**, 1050–1060.

114. Lee, H., Lee, K. D., Pyo, K. B., Park, S. Y., and Lee, H. (2010). Catechol-grafted poly(ethylene glycol) for PEGylation on versatile substrates, *Langmuir*, **26**, 3790–3793.

115. Sever, M. J., and Wilker, J. J. (2004). Visible absorption spectra of metal-catecholate and metal-tironate complexes, *Dalton Trans.*, 1061–1072.

116. Rodriguez, R., Blesa, M. A., and Regazzoni, A. E. (1996). Surface complexation at the TiO_2 (anatase) aqueous solution interface: Chemisorption of catechol, *J. Colloid Interf. Sci*, **177**, 122–131.

117. Martin, S. T., Kesselman, J. M., Park, D. S., Lewis, N. S., and Hoffmann, M. R. (1996). Surface structures of 4-chlorocatechol adsorbed on titanium dioxide, *Environ. Sci. Technol.*, **30**, 2535–2542.

118. Shultz, M. D., Reveles, J. U., Khanna, S. N., and Carpenter, E. E. (2007). Reactive nature of dopamine as a surface functionalization agent in iron oxide nanoparticles, *J. Am. Chem. Soc.*, **129**, 2482–2487.

119. Xu, C. J., Xu, K. M., Gu, H. W., Zheng, R. K., Liu, H., Zhang, X. X., Guo, Z. H., and Xu, B. (2004). Dopamine as a robust anchor to immobilize functional molecules on the iron oxide shell of magnetic nanoparticles, *J. Am. Chem. Soc.*, **126**, 9938–9939.

120. Bin Na, H., Palui, G., Rosenberg, J. T., Ji, X., Grant, S. C., and Mattoussi, H. (2012). Multidentate catechol-based polyethylene glycol oligomers provide enhanced stability and biocompatibility to iron oxide nanoparticies, *Acs Nano*, **6**, 389–399.

121. Lee, Y. H., Lee, H., Kim, Y. B., Kim, J. Y., Hyeon, T., Park, H., Messersmith, P. B., and Park, T. G. (2008). Bioinspired surface immobilization of hyaluronic acid on monodisperse magnetite nanocrystals for targeted cancer imaging, *Adv. Mater.*, **20**, 4154–4157.

122. Amstad, E., Gillich, T., Bilecka, I., Textor, M., and Reimhult, E. (2009). Ultrastable iron oxide nanoparticle colloidal suspensions using dispersants with Catechol-derived anchor groups, *Nano Lett.*, **9**, 4042–4048.

123. Long, Y. H., Wu, J. J., Wang, H., Zhang, X. L., Zhao, N., and Xu, J. A. (2011). Rapid sintering of silver nanoparticles in an electrolyte solution at room temperature and its application to fabricate conductive silver films using polydopamine as adhesive layers, *J. Mater. Chem.*, **21**, 4875–4881.

124. Fei, B., Qian, B. T., Yang, Z. Y., Wang, R. H., Liu, W. C., Mak, C. L., and Xin, J. H. (2008). Coating carbon nanotubes by spontaneous oxidative polymerization of dopamine, *Carbon*, **46**, 1795–1797.

125. Xu, L. Q., Yang, W. J., Neoh, K. G., Kang, E. T., and Fu, G. D. (2010). Dopamine-induced reduction and functionalization of Graphene oxide nanosheets, *Macromolecules*, **43**, 8336–8339.

126. Kang, S. M., Park, S., Kim, D., Park, S. Y., Ruoff, R. S., and Lee, H. (2011). Simultaneous reduction and surface functionalization of graphene oxide by mussel-inspired chemistry, *Adv. Funct. Mater.*, **21**, 108–112.

127. Kaminska, I., Das, M. R., Coffinier, Y., Niedziolka-Jonsson, J., Sobczak, J., Woisel, P., Lyskawa, J., Opallo, M., Boukherroub, R., and Szunerits, S. (2012). Reduction and functionalization of graphene xxide sheets using biomimetic dopamine derivatives in one step, *ACS Appl. Mater. Interfaces*, **4**, 1016–1020.

128. Liu, R., Mahurin, S. M., Li, C., Unocic, R. R., Idrobo, J. C., Gao, H. J., Pennycook, S. J., and Dai, S. (2011). Dopamine as a carbon source: the controlled synthesis of hollow carbon spheres and yolk-structured carbon nanocomposites, *Angew. Chem. Int. Edit.*, **50**, 6799–6802.

Chapter 4

Gecko Adhesion

Joseph C. Cremaldi, Kejia Jin, and Noshir S. Pesika

Department of Chemical and Biomolecular Engineering, Tulane University, 333 Lindy Boggs Building, New Orleans, LA 70118-5674, USA

npesika@tulane.ed

In this chapter, we review the experimental and theoretical research over the last 2 decades that has provided insights into the mechanisms involved in the gecko adhesion system. Theoretically models developed to explain the numerous properties of the gecko adhesion pads, including strong adhesion, anisotropic adhesion, self-cleaning, shear-induced adhesion, and adhesion under various environmental conditions and to surfaces of different roughness, are discussed. Experimental results from various research groups are also presented consistent with the theoretical analyses, e.g., the peel zone model, which predicts the adhesion force as a function of the peel angle and explains the anisotropic adhesion in geckos. Although researchers in field have managed to explain several of the unique properties of the gecko adhesion system, there still remain open questions, which need to be answered before fully understanding the gecko adhesion system. These answers will be crucial for researchers seeking to develop a true gecko-like dry adhesive.

Bioadhesion and Biomimetics: From Nature to Applications
Edited by Havazelet Bianco-Peled and Maya Davidovich-Pinhas

ISBN 978-981-4463-98-0 (Hardcover), 978-981-4463-99-7 (eBook)
www.panstanford.com

4.1 Introduction

The ability of geckos to run along vertical and inverted surfaces has been a source of curiosity for scientists for a very long time. Over the years, people have theorized that this ability has been a product of various properties of a gecko's foot, or pad. These theories have included the use of glue-like secretions, suction [11,13], microscale interlocking [25], electrostatics [44], capillary adhesion [8,10,25], and van der Waals forces [36]. As the capabilities of microscopy have increased along with lowered costs over the years, more labs acquired the resources to try and tackle this scientific curiosity. Likewise, the theorized explanations of the gecko's adhesive pads have been slowly ruled out over the years due to various shortcomings; performance in a vacuum ruled out suction [10], lack of glands on the pads ruled out wet adhesion through capillary action and glues [8,10,25], irregular surface-scaling of the spatulae ruled out microscale interlocking [24], and the non-effect of X-rays on the adhesive capability ruled out electrostatic interactions [10]. Finally, in 2000, Keller Autumn and his team reported conclusively that van der Waals forces are the source of adhesion in their article titled "Adhesive Force of a Single Gecko Foot-Hair [5]." Along with this essential publication, scientists have been unraveling the complex mechanisms and structures that grant geckos their exceptional abilities.

In addition to the adhesive abilities of the gecko, scientists have uncovered other properties within the gecko adhesive structures that allow them to effortlessly and quickly climb on a variety of surfaces; anisotropic (directional) forces, force generation in relation to preloading, ease of detachment, material-independence of attachment, non-self-adherence, and the self-cleaning mechanism. In order to accommodate these various properties, it has been found that several of the gecko's biological systems have been incorporated into a complex and unique hierarchical system [35–41]. These various biological systems include the integration of the integumentary, circulatory, musculoskeletal, sensory, and connective tissues [37,43]. For the purposes of this chapter's discussion on the gecko, the mechanical properties of the gecko as a whole will be analyzed without going into the biological functions underlying those abilities.

4.2 The Gecko Pad

In order to begin to understand all of the various abilities of the gecko as they pertain to its locomotion, one must first understand the basic makeup of a Tokay gecko's foot. A gecko's foot is often referred to as a pad, and will be so in the context of this text. The gecko pad is a hierarchical structure that exists on a macro-, meso-, micro-, and nano- scale as shown in Fig. 4.1.

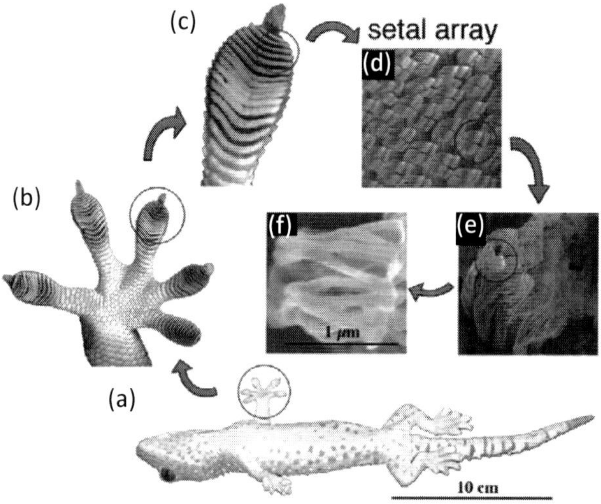

Figure 4.1 Hierarchical breakdown of the Tokay gecko foot (pad) [47]: (a) Optical image of an inverted gecko showing the location of the gripping pad on a toe. (b) Higher magnification image of the pad placement on the underside of the foot. (c) Lamellar ridges composed of soft tissues that serve as a more elastic underlay to the gecko setae. (d) Setae structures attached to the lamellar ridges. (e) Setae tips. (f) Spatulas.

4.2.1 Hierarchical Tiers

Starting at the largest scale, Fig. 4.1a–f shows the different levels of hierarchy of the gecko adhesive system. While 60% of the 1450 gecko species exhibit the ability to move on walls and inverted surfaces [12], the model organism most used in gecko adhesion studies is the Tokay Gecko (*gekko gecko*) due to its larger size and general abundance. A general comparison of contact areas among various gecko species was completed by Irschick [21] and displays the large relative contact area that Tokay geckos utilize. The Tokay Gecko is the second largest gecko species and ranges in length from 11 to 20 inches for males and approximately 7–19 inches for females. A large, full-grown Tokay gecko weighs 50–100 g [7,21]. Macroscale features of a gecko include four legs with five toes each. At the mesoscale level, the underside of each toe has a series of approximately 20 ridges called lamellae. They are spaced approximately 1–2 mm apart and provide an elastic tissue underlayer. On the microscopic level, on each lamellar ridge exists a dense packing of β-keratin hair-like setae. Each seta is approximately 30 to 125 microns in length and 5 to 10 microns in diameter [34,36,49]. The last, nano-scale tier of the structuring is the tips of the setae. The end of each setae splits into approximately 100–1000 smaller spatulas 50–250 nanometers in length. The triangular ends of the spatulas are larger and flatter, about 10 nm thick and about 200 nm wide [36,49]. These tiers result in hundreds of millions to billions of spatulas with a total macroscopic contact area of 200–250 mm^2 [21,48]. Table 4.1 provides a breakdown of the relationships in terms of number and areas of the various tiers of the gecko pad. It is worth noting that a recent discovery shows that phospholipids are deposited in the wake of gecko movement, leaving a hydrophobic footprint in their wake [17]. The release of these lipids does not act as glue in any way, but may play a yet unknown part in the mechanics of gecko adhesion.

4.2.2 Van der Waals Forces

With the hierarchical structure established, we return to the source of adhesion, i.e., the relatively weak attractive van der Waals force based on the proximity of any two molecules. These forces occur between the permanent or induced (temporary) dipoles that exist in molecules: dipole-dipole, induced dipole-dipole, induced dipole-induced dipole. In addition to the general weakness of the force, it decreases rapidly as the separation distance increases and is effective only in the atomic interaction range of angstroms. Due to the weak interactions and miniscule distances necessary, in order for this force to be a factor in adhesion, large areas of intimate contact between two surfaces must be achieved. The van der Waals force per unit area between two flat, parallel surfaces can be expressed as follows [47]:

Table 4.1 Breakdown of the numbers and areas covered by a gecko pad

	Pre gecko	Per foot	Per toe	Per lamella	Per seta	Per spatula	Characteristics
Gecko	1	N/A	N/A	N/A	N/A	N/A	Tokay Gecko (Gekko Geckoj) used as a model organism: approximately 11 to 20 inches and 150 to 400 grams
Foot	4	1	N/A	N/A	N/A	N/A	4 legs
Toe	20	5	1	N/A	N/A	N/A	5 toes
Lamella	400	100	20	1	N/A	N/A	20 ridges per toe composing an elastic tissue under-layer and spaced approximately 1–2 mm apart
Seta	5,600,000	1,400,000	280,000	5,000-14,400 (14000)	1	N/A	β-keratin hair-like structures 30 to 125 microns in length and 5 to 10 microns in diameter
Spatula	2,800,000,000	700,000,000	140,000,000	7,000,000	100-1,000 500	1	50–250 nanometers in length with triangular ends (larger and flatter) approximately 10 nm thick and about 200 nm wide
Total Area [mm^2]	440.00	110.00	22.00	1.10	1.10E-04	2.20E-07	

Note: Note that different sources provide various values [5,7,16,19,34, 36, 49].

$$F_{\mathrm{vdW}} = \frac{A}{6\pi D^3} \tag{4.1}$$

In Eq. 4.1, F_{vdW} is the van der Waals force, D is the gap distance between the two surfaces, and A is the material-dependent Hamaker constant.

4.2.3 Roughness Considerations

When people envision intimate contact, they envision two very flat surfaces in contact with each other. Unfortunately, nature does not subscribe to such a notion and most surfaces, even those that appear flat macroscopically, have a certain degree of roughness on the nano- and micro-scales. It is straightforward to see how the bending of legs and toes allows the gecko to get good contact on a macroscopic level (grappling a branch, etc.). The unique design of the gecko, however, is where it deviates from most other organisms in that its toes continue to provide structural methods of dealing with roughness down to the nanoscale levels through the meso-, micro-, and nanoscale features. Figure 4.2 shows the actual roughness of three common surfaces and displays how the hierarchical structures effectively create intimate contact.

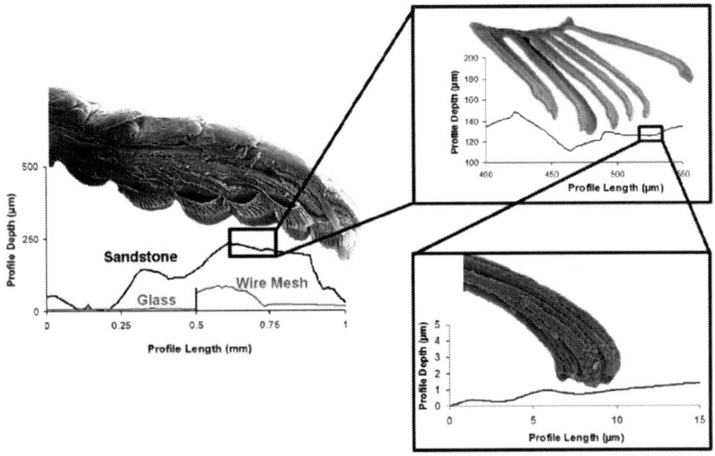

Figure 4.2 Hierarchical image of how the different levels (sizes) of the gecko pad can deal with roughness at different length scales [43]. By matching smaller features with smaller roughness scales, the pad effectively flattens the surface.

In looking at the way various adhesives deal with roughness, Persson introduced an effective material modulus that is essential in dealing with rough surfaces [28,29]. Although β-keratin has a large modulus of ~2 GPa, this effective material modulus explains how fibers (setae and spatulae) allow such a hard material to introduce elasticity and the ability to conform to uneven surfaces. The effective elastic modulus of the setae is much lower, at approximately 100 kPa [6], than that of bulk β-keratin due to the microstructure of the setae. This value moves the gecko pad into the Dahlquist criterion range for tack.

4.2.4 Force Generation and the Autumn Experiment

Just how much adhesive (perpendicular to contact) and frictional (parallel to contact) forces does a gecko generate? The structure of the gecko pad had been determined over the course of the 20th century and measurements had been taken on the shear adhesion (friction force) capabilities of the entire gecko pad [10,34,36]. In 1996 Irschick et al. reported a force of 20 N for two feet of a Tokay Gecko; approximately 200–220 mm^2 and three million setae [21]. In 2001, however, the Autumn group determined the forces associated with a single seta 200 μN of friction force and 40 μN of adhesive force and confirmed the van der Waals mechanism of adhesion [5]. The experimental setup utilized a single seta attached to a micro-electromechanical system (MEMS) consisting of a piezo-resistant cantilever. The testing surface consisted of a flattened aluminum wire in which the displacement of the wire could be measured through compound optical microscopy and image analysis software.

4.2.5 Force Results

Before discussing the test results, it is important to note that the orientation of the seta was tracked optically during the force measurements. The gecko creates forces by placing its foot on the substrate (preload) and then bringing the toes inwards toward the center of the foot (gripping direction) as seen in Fig. 4.3a.

The first test consisted of a parallel force (friction) measurement utilizing a known perpendicular preload. Once the preload was applied, the seta was moved in the parallel gripping direction at a constant velocity and the resulting friction and adhesion forces were measured (Fig. 4.3b). Based on a general 100 mm^2 contact area and the fact that there exist approximately 5000 setae per mm^2, a resultant force of 20 μN could be expected for each seta [21]. Due to the fact that in most cases all of the setae did not contact the surface simultaneously, a slightly larger number may have been possible. However, the experimental results gave a value almost an order of magnitude higher at 200 μN of friction force and 40 μN of adhesive force. These results were obtained when a small preload was applied and the seta was allowed to slide over the surface a small distance of ~5 μm. It is important to remember that at 10 N per foot and four feet the gecko was expected to already support an adhesive weight of a little over 4 kg, already a high number to support an organism that weighs between 0.5 to 0.1 kg. The new results meant that at 100% contact, 100 mm^2 of contact area could be expected to achieve 40 kg of friction force (referred to as shear adhesion in the referenced article).

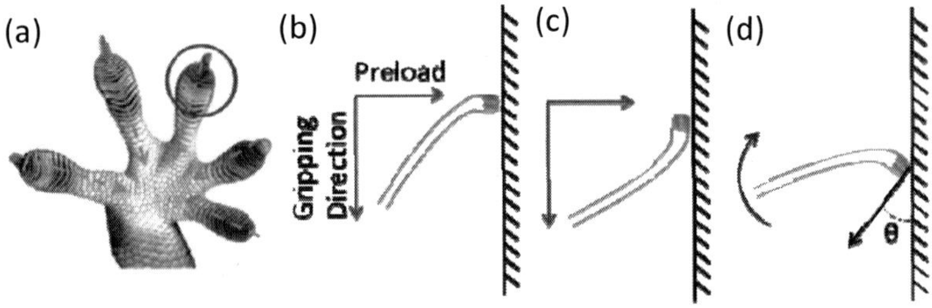

Figure 4.3 (a) Optical image of a gecko foot. The arrows point toward the gripping direction [5]. (b) Schematic illustration of the tribological measurements in the gripping direction. (c) Schematic illustration of the tribological measurements proving the effect of the spatulas in force generation. (d) Schematic illustration of the experiment used to determine the critical angle needed for no adhesion (i.e., detachment).

The second test simply reversed the orientation of the seta so that the spatulas faced away from the substrate and then repeated the first test with varying preload (Fig. 4.3c). The resulting forces were 100 times smaller and showed that unless the spatulas contacted the substrate, the seta behaved in a similar fashion to a material with a coefficient of friction of ~0.25 and force generation varied linearly with the preload. This test served to confirm that the spatulas play a direct role in creating intimate contact required to generate the van der Waals attractive forces.

The third and last test varied the angle θ at which the seta (and therefore spatulae) contacted and finally detached from the surface as can be seen in Fig. 4.3d. The data showed that there exists a critical angle of approximately 30° above which the seta no longer causes adhesion to the substrate. This test offered an insight into the basic mechanism that would account for the ease of movement and ability to quickly detach their feet in approximately 15 ms [5], required for rapid locomotion.

4.2.6 Comparison with Whole Pad and Single Spatula Measurements

While Autumn's tests provided direct evidence for the van der Waals mechanism of adhesion, tests had previously been completed on whole foot-pad measurements. It has been mentioned previously that Irschick reported 20 N of force for two feet [21]. This seemingly low number shows how efficiency plays into the total force generation of the gecko. While the gecko theoretically generates enough force to support its body mass 100-fold, the reverse statement shows that only 1% of the spatulae need to be in contact for this mechanism to work. This is the much more realistic case that would be present in nature during a gecko's normal everyday movement over rough surfaces. Additionally, in the time since the reporting of this discovery, one research group found a method to take measurements on a single spatula. In 2005, Huber et al. reported tests on force measurements of a single spatula using an atomic for microscope (AFM) cantilever with advanced sample preparation techniques [19]. There resulting force of 10 nN per spatula is approximately four times less than those found in the (Autumn) experiments outlined above. This discrepancy will be covered after a discussion of how the method of testing affected the two apparently different results.

4.3 Attachment and Detachment

At this point, a more detailed discussion of the setae positioning in relation to the surface as well as the mechanism by which a gecko moves its toes to attach and detach from a surface is necessary. The gecko's toes spread from the leg in five directions radially from the center. With each step, the gecko utilizes two of its opposite leg sets, either back right/forward left or back left/forward right, in order to pull inwards toward the gecko's center, known as the Y configuration [30]. With a small downward preloading force and the inward motion to create the parallel force from sliding in order to obtain the 5 μm displacement, the gecko generates the large adhesion and friction forces associated with each foot.

In terms of detachment, Autumn's group found that little to no measurable forces were recorded after the setal angle reached 30° [3]. The gecko pad (Fig. 4.4a) goes through digital hyperextension [36,39], which causes the toes to roll upward and backward (Fig. 4.4b) toward the legs. Figure 4.4c provides a schematic illustration of the rolling mechanism.

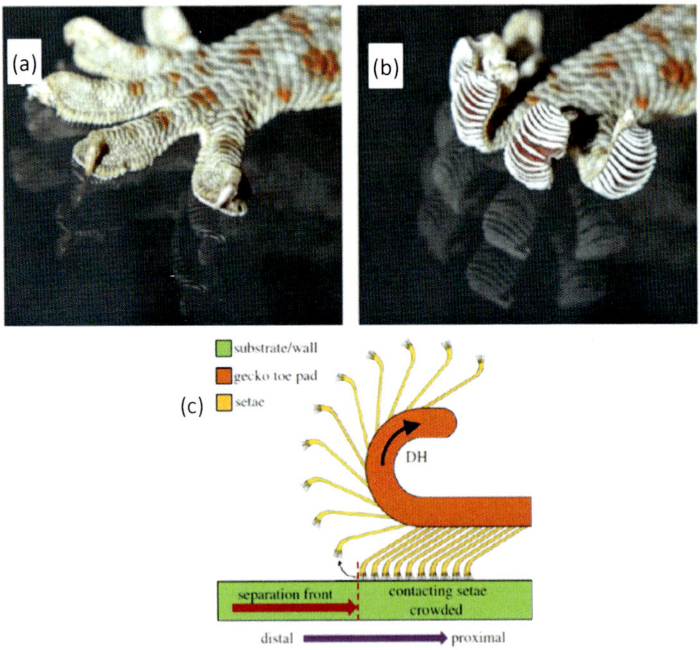

Figure 4.4 (a) Optical image showing a gecko's foot attached to a substrate. (b) Optical image showing a gecko's foot with the digits hyperextended. (c) Schematic illustration of how the hyperextension motion introduces a mechanism by which the setae "roll" off the surface and create large peel angles [18].

In doing so, the toes cause the setae to reach the 30° threshold at which the detachment forces become very small almost to the point of being negligible. The reason this occurs is that during this rolling motion, only a tiny fraction of the setae are being detached at any one time. The "rolling" motion of the toes allows for a modified force model based on peeling tape off a substrate (Fig. 4.5). A peel force is generated by the rolling action of the gecko toe and can be described mathematically. Figure 4.5 shows the development of the actual gecko spatulae (Fig. 4.5a–c) into a tape-peeling model (Fig. 4.5d,e).

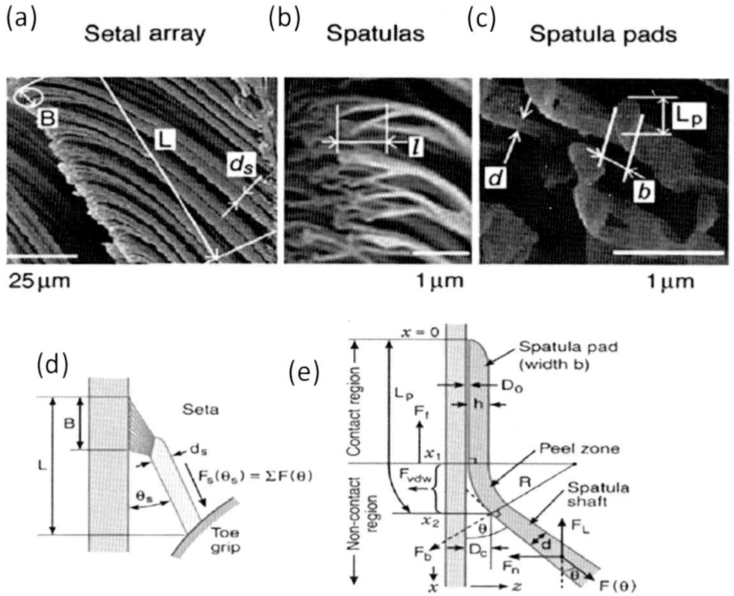

Figure 4.5 (a, b, c) SEM images of the setae and spatulae. (c, d) Schematic illustration of the peel-zone Model [30].

4.3.1 The Peel-Zone Model

A tape peeling model first introduced by Rivlin [33] and later modified by Kendall to include the elastic energy of the tape substrate serves as a starting point in the analysis [22]. The model aptly describes the detachment of elastically backed tapes to rigid surfaces, where the gecko detachment can be viewed as the elastic tape in this model.

$$\frac{F}{b} = \frac{\gamma}{(1-\cos\theta)} + \text{elastic energy} \tag{4.2}$$

In the modified Kendall Equation for tape peeling (Eq. 4.2), *F* is the peel force, *b* is the characteristic length, γ is the energy of adhesion, and θ is the peel angle. If solved for the peel force, the amount of force required to peel the tape to a new distance is given when the peel angle is held constant. Previous studies considering the loss of energy due to slip have been carried out without consideration of the peel angle [1,26]. The Kendall model describes the macroscopic, bundled effects of the peel-zone rather than further analysis of a single gecko spatula [47].

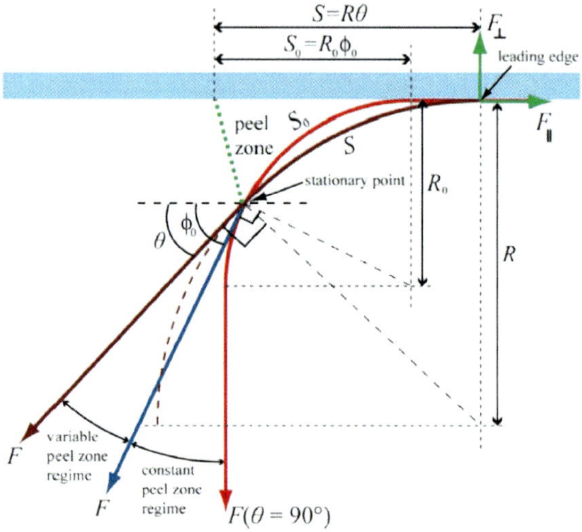

Figure 4.6 Schematic illustration of the peel-zone model with the different modes of peeling that can occur. This model shows two distinct cases; peel angles between Case I and Case II known as the constant-peel regime and the angles between Case II and Case III known as the variable-peel regime [30].

In 2006, Tian et al. introduced the peel-zone model, which utilizes a force balance at the spatula level and adds a peel angle-dependent multiplier to the Kendall Model. The peel zone is the area between the two surfaces in which they are splitting (bifurcating) and it is characterized by cavitation and fibrillation (in the case of adhesive tapes) or where van der Waals forces are predominant (in the case of gecko spatulas). Using Fig. 4.6, one can more easily see that the peel zone (S) represents the area where the separation of the tape (or gecko pad) from the substrate takes place and where the spatulae are still interacting with the surface. The zone exists between the "pin point" (solely based on parallel friction forces) at the right hand side of the length, *S*, to the point of the last fibril in contact with the substrate surface on the left. At this point, certain assumptions are taken into account. First, we assume a circular curvature in dealing with the backing of the tape in the peel zone region. This allows one to more directly translate the peel zone into arc lengths for calculations and modeling. Based on this radius of curvature *R*, the peel zone length becomes directly proportional to the arc length (i.e., $R\theta$) and an average distribution of forces over that length is utilized. Second, it is assumed that the tape consists of a large modulus, or does not stretch any considerable amount along its length. In terms of reality, the β-keratin composition of the setae is well represented in that the elastic modulus is approximately 2 GPa. Again referring to Fig. 4.6, one can see that as the peel angle reduces, the larger the parallel component of the force gets opposed through the parallel friction force. Similarly, as the peel angle increases more of the peel force is translated into the perpendicular adhesive component.

4.3.2 Peel Modes

This model shows two distinct cases; peel angles between Case I and Case II known as the constant-peel regime and the angles between Case II and Case III known as the variable-peel regime. In the constant peel regime, the length of the peel zone as well as the tape substrate curvature remains constant while the peel angle changes. In the variable case these two amounts change with a changing peel angle. In either case and utilizing the assumption of a large stretch modulus, one can say that the active front of the peel zone remains at a constant normal distance from the backing of the tape (the last filament attached can stretch a certain amount before it detaches and the next closest assumes that position). Utilizing geometric relationships that are beyond the scope of this chapter, the variable peel-zone detachment can be shown to reduce to

$$\frac{F(\theta,\phi_0)}{b}=\gamma\left[\frac{\theta}{\phi_0}\right]\left[\frac{1-\cos\phi_0}{1-\cos\theta}\right]\left[\frac{1}{\sin\theta}\right] \tag{4.3}$$

In Eq. 4.3, F is the total peel force and b is the width of the tape (spatula width in the gecko model). The characteristic force per unit width (F/b) is a function of the peel angle θ and ϕ_0, the angle that the last active filament creates with the substrate material as shown in Fig. 4.6. The angle ϕ_0 is an intrinsic material property that is based on the material properties of both the substrate and the adhesive, and is determined experimentally. The difference between this model and the Kendall model is that the increase in the length of the peel zone, S, is accounted for in this model, and shows that the forces predicted by the Kendall model will always be too large based on the angle-dependent multiplier.

In relation to the actual gecko, this model takes into account the β-keratin's large elastic modulus and typically small peel angles. The scale of the individual setae is particularly difficult to study, however, meaning that modeling the individual setae introduces large errors. This experiment on a single spatula has been conducted by Huber et al. They found the maximum pull-off force to be one quarter of that found by Autumn, or 10 nN per spatula [19]. However, based on the overall friction and well-documented number of spatula per seta, one can deduce that this number is too low. This particular experiment is utilized because the slight change in methods between the two experiments offers a glimpse into the nanoscale mechanism of peel-off. Huber's experiment measured the peel off force by pulling perpendicular to the surface, similar to the common T-peel test. This is analogous to keeping the peel angle (θ) constant at 90°. On the other hand, Autumn's test on a single seta sheared the seta at an angle smaller than 90°, thus resulting in larger adhesive forces consistent with the peel-zone model. Now that the mechanism of peel-off has been explained, it is more clearly seen how the gecko can accomplish this biologically. By bending its toes away from the plane of contact, digital hyperextension, the spatulas are exposed to larger peel angles until the threshold is met and the adhesive forces per area drop to almost zero. This same hyperextension brings us to the next very important quality of a gecko's movement that keeps it functional.

4.4 Self-Cleaning Mechanism

One may wonder how a foot pad that can attach to almost any surface could possibly remain clean. This self-cleaning ability is due to an energetic imbalance of particulate that may attach to a gecko's pad. Furthermore, experiments in which this hyperextension has been prevented during gecko movement show that this ability further amplifies the ability to self-clean by almost twofold due to a dynamic dislodging of the spatulas during the rolling off mechanism [18].

Without digital hyperextension, self-cleaning occurs with each step taken by the gecko. The reason is that the contact area between the spatulas and a typical dust particle is not equal to the forces being generated by the particle and the substrate (wall). Therefore, the dust particle will prefer to adhere to the wall in preference to the spatula [2,15] as can be seen in Fig. 4.7. This has been confirmed by Hu et al. in an experiment in which movement-prohibiting boots were placed on the gecko's feet to restrict movement [18]. In doing so, they found that the mechanism of detachment changed from a rolling off peel-zone format to a crack propagation form moving from the proximal end of the toe to the distal end.

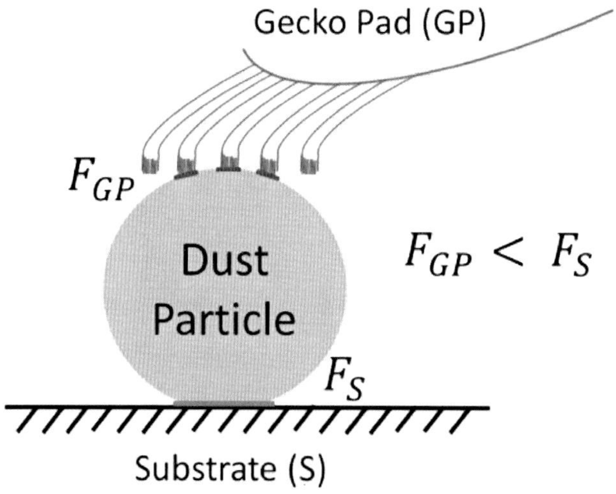

Figure 4.7 Schematic illustration of the unbalanced force generation favoring the substrate over the gecko pad, enabling the gecko's ability to self-clean.

This is opposite to the normal distal rolling associated with digital hyperextension. The gecko pads cleaned themselves regardless of the rolling motion. However, once they took off the prohibitive boots, they found that the time to achieve the same amount of pad cleansing was almost halved. Their explanation is twofold. First, using the rolling motion, the number of setae decreases as the disattachment moves across the dirt particle. So, in effect the particle remains pinned to the substrate by an ever decreasing amount of setae. In addition, Hu et al. suggest that there exists a dynamic component to the mechanism that removes the particulate from the pads. While the spatulas are rolling during digital hyperextension, large inertial forces are generated at the tips during disattachment. In essence, a buildup of elastic energy [14] in the setae and spatulas is suddenly released as they rotate upwards. When combined with the very small contact area, this large force is enough to propel the particulate off the pad [18]. These mechanisms allow the gecko to generate large friction and adhesive forces to substrates on a large scale while simultaneously removing dirt from the pads during locomotion.

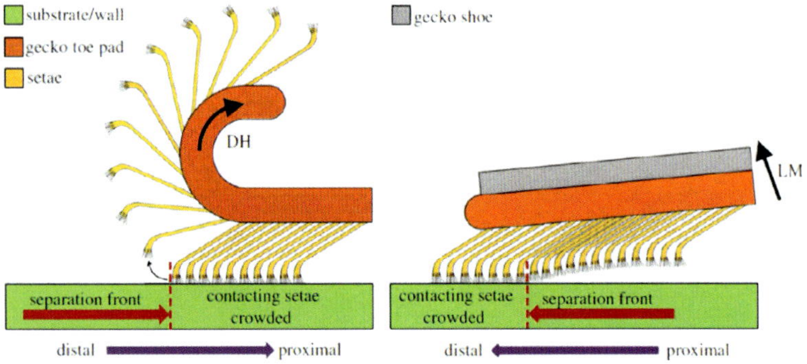

Figure 4.8 The test setup used to change the gecko movement from igital hyperextension (DH) to a lift-off mechanism (LM) and change the direction of the crack propegation front. This change showed the effects of dynamic self-cleaning mechanism by Hu et al. [18].

4.5 Non-Adhesive Unloaded State and Non-Self-Adherence

Another factor is the unloaded contact area, which has been shown to be approximately 6% of total area [4]. The seemingly excessive capabilities of the gecko to adhere can be explained by the need for quick adherence in short distances when in an emergency situation such as falling [2]. A side effect of this small contact area is that the gecko's feet are superhydrophobic when unloaded due to the hydrophobicity of the β-keratin and the structural nature of the setal array. If the surface of the pad were to come into contact with water, the lotus effect would cause an easy washing mechanism to remove debris. The amount of time that a gecko spends in contact with water is unknown, however. While mentioning the non-adhesive unloaded state of the gecko pads, another two straightforward properties can be quickly mentioned. First, gecko pads will not self-adhere due to the extremely small areas of contact between the two pads [4]. Second, the spatulas will not come together because the modulus of the β-keratin is high [4]. Essentially it amounts to trying to get two incredibly small springs to stay together when they are pulling apart and the van der Waals forces cannot overcome the desire of the spatulas to separate as shown in Fig. 4.9.

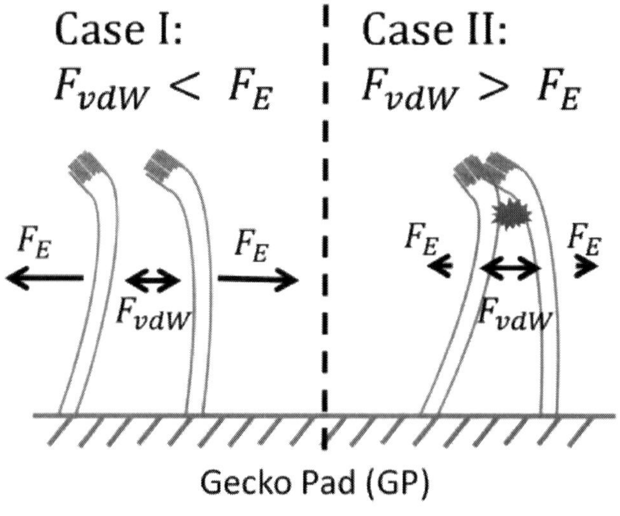

Figure 4.9 The modulus of the β-keratin generates a large elastic force (F_E). In Case I the elastic bending forces are large enough to overcome the attractive van der Waals forces (F_{vdW}). In Case II the van der Waals force (F_{vdW}) overcomes the elastic bending forces (F_E), causing the two setae to come together.

4.6 Substrates and the Environment

Now that the mechanism by which a gecko can adhere to and detach from surfaces has been discussed, it is time to discuss other relevant aspects pertinent to gecko adhesion, i.e., the surface being attached to and the surrounding environment. By utilizing van der Waals forces, one might think that the mechanism is substrate independent. However, while geckos have shown the ability to attach to both hydrophilic and hydrophobic surfaces with ease as would be expected, it has been widely shown that geckos cannot transverse a Teflon surface or that the incline of the surface may play a large role in the translation of a surface [42]. A similar environmental consideration has been researched on the temperature and humidity effects on the adhesive and frictional capabilities of geckos. Many studies have proved that the relative humidity [20] and temperature play a large role in the effectiveness of the gecko's adhesion abilities. Researchers are continuing to look at the extent of capillary effects on the gecko's adhesion abilities [9,23,46]. Results revolving around the use of both hydrophobic and hydrophilic surfaces have shown the results to be surface independent, meaning that humidity and temperature play a role in the gecko itself [27]. Disregarding the effects of biological factors, β-keratin softens and becomes more pliable with the increase in humidity [32], thereby increasing the gecko pad's ability to conform to the surface as a whole. Because the relative humidity is so closely tied to the temperature, most consider this to be the explanation for any variation in temperature testing. This recent development is continuing to be explored to explain different modes of dynamic friction (i.e., friction as a function of shearing) in the gecko pad [31]. At low shearing velocities the friction force of the adhesive pads are very dependent on the material properties that can change with relative humidity. As the velocity increases, the friction force becomes material and humidity independent. At the extreme end of environmental testing are the capabilities of gecko pads to perform underwater on both hydrophobic and hydrophilic surfaces. Tests show that in underwater conditions, gecko pads retain similar characteristics to their dry counterparts when dealing with hydrophobic and weakly hydrophobic (weakly hydrophilic) surfaces. However, when dealing with hydrophilic surfaces, the submerged samples provide for a quarter to a third as much adhesion [45]. Additionally, when tested on a Teflon (PTFE) surface that traditionally is almost non-adhesive to the gecko, the submerged sample provided for four to five times greater adhesion than its dry (Teflon) counterpart. Tests such as these show that there is still a great deal to learn about gecko adhesion and may offer insight into the role of the recently discovered anti-wetting phospholipids mentioned previously [17].

4.7 Conclusion

In this chapter, the basics of gecko adhesion have been covered. By utilizing a sophisticated hierarchical structuring, the gecko's foot can conform to roughness down a nano-scale level. In doing so, the gecko achieves intimate contact with billions of spatulas, where the net force achieves a large overall adhesion and friction capabilities through van der Waals forces. Being a recent discovery, more exact measurements and models are being developed constantly, with the peel-zone model a good current representation of the mechanism. By using the peel-zone model, the biological mechanism of hyperextension by which a gecko achieves the large peel-angles needed to create almost zero detachment forces has been well-documented. The substrates and environmental effects are now being solidified. More particularly, the dual temperature/relative humidity effects on the materials are being researched and initial data shows that it increases compliance thereby further increasing the contact of the two surfaces. With the extraordinary abilities that the gecko's adhesive pads exhibit, creating a synthetic version has become a booming research area. Several properties are advantageous over the typical pressure sensitive adhesives including anisotropy, self-cleaning abilities, and surface independence to name a few.

References

1. Amouroux, N., Petit, J. R. M., and Léger, L. (2001). Role of interfacial resistance to shear stress on adhesive peel strength. *Langmuir.*, 6510–6517.
2. Autumn, K. (2006). How gecko toes stick: The powerful, fantastic adhesive used by geckos is made of nanoscale hairs that engage tiny forces, inspiring envy among human imitators. *Am. Sci.*, 124–132.
3. Autumn, K., Dittmore, A., Santos, D., Spenko, M., and Cutkosky, M. (2006). Frictional adhesion: A new angle on gecko attachment. *J. Exp. Biol.*, 3569–3579.

4. Autumn, K., and Hansen, W. (2006). Ultrahydrophobicity indicates a non-adhesive default state in gecko setae. *J. Comp. Physiol.*, 1205–1212.

5. Autumn, K., Liang, Y. A., Hsieh, S. T., Zesch, W., Chan, W. P., Kenny, T. W., Fearing, R., and Full, R. J. (2000). Adhesive force of a single gecko foot-hair. *Nature*, 681–685.

6. Autumn, K., Majidi, C., Groff, R. E., Dittmore, A., and Fearing, R. (2006). Effective elastic modulus of isolated gecko setal arrays. *J. Exp. Biol.*, 3558–3568.

7. Autumn, K., and Peatie, A. M. (2002). Mechanisms of adhesion in geckos. *Int. Comp. Biol.*, 1081–1090.

8. Bellairs, A. (1970). *The Life of Reptiles*, vols. 1 and 2. New York, Universe Books.

9. Bhushan, B. (2003). Adhesion and stiction: Mechanisms, measurement techniques,and methods for reduction. *J. Vac. Sci. Technol. B.*, 2262–2296.

10. Dellit, W.-D. (1934). Zur Anatomie und Physiologie der Geckozehe. *Jenaische Zeitschrift fuer Naturwissenschaft.*, 613–656.

11. Gadow, H. (1901). *Amphibians and Reptiles*. London, McMillan and Company.

12. Gamble, T., Greenbaum, E., Jackman, T. R., Russel, A. P., and Bauer, A. M. (2012). Repeated origin and loss of adhesive toepads in geckos. *PLoS ONE*, e39429.

13. Gennaro, J. F. (1969). The gecko grip. *Nat. His.*, 36–43.

14. Gravish, N., Wilkinson, M., and Autumn, K. (2008). Frictional and elastic energy in gecko adhesive detachment. *J. R. Soc. Interface*, 339–348.

15. Hansen, W. R., and Autumn, K. (2005). Evidence for self-cleaning in gecko setae. *Proc. Nat. Acad. Sci.*, 385–389.

16. Hiller, U. (1975). Comparative studies on the functional morphology of two gekkonid lizards. *J. Bombay Nat. Hist. Soc.*, 278–282.

17. Hsu, P. Y., Ge, L., Li, X., Stark, A. Y., Wesdemiotis, C., Niewiarowski, P. H., and Dhinojwala, A. (2012). Direct evidence of phospholipids in gecko footprints and spatula-substrate contact interface detected using surface-sensitive spectroscopy. *J. R. Soc. Interface*, 657–664.

18. Hu, S., Lopez, S., Niewiarowsk, P. H., and Xia, Z. (2012). Dynamic self-cleaning in gecko setae via digital hyperextension. *J. R. Soc. Interface*, 2781–2790.

19. Huber, G., Gorb, S. N., Spolenak, R., and Arzt, E. (2005). Resolving the nanoscale adheion of individual gecko spatulae by atomic force microscopy. *Biol. Lett.*, 2–4.
20. Huber, G., Mantz, H., Spolenak, R., Mecke, K., Jacobs, K., Gorb, S. N., and Arzt, E. (2005). Evidence for capillarity contributions to gecko adhesion from single spatula nanomechanical measurements. *Proc. Nat. Acad. Sci.*, 16293–16296.
21. Irschick, D. J., Austin, C. C., Petren, K., Fisher, R. N., Losos, J. B., and Ellers, O. (1996). A comparative analysis of clinging ability among pad-bearing lizards. *Biol. J. Linnean Soc.*, 21–35.
22. Kendall, K. (1975). Thin-film peeling: The elastic term. *J. Phys. D: Appl. Phys.*, 1449–1452.
23. Kim, T. W., and Bhushan, B. (2008). The adhesion model considering capillarity for gecko attachment system. *J. R. Soc. Interface*, 319–327.
24. Maderson, P. F. A. (1964). Keratinized epidermal derivatives as an aid to climbing in gekkonid lizards. *Nature*, 780–781.
25. Mahendra, B. C. (1941). Contributions to the bionomics, anatomy, reproduction and development of the Indian house-gecko, hemidactylus flaviviridis ruppel. *Proc. Indian Acad. Sci.*, 288–306.
26. Newby, B.-M. Z., Chaudhury, M. K., and Brown, H. R. (1995). Macroscopic evidence of the effect of interfacial slippage on adhesion. *Science*, 1407–1409.
27. Niewiarowsk, P. H., Lopez, S., Ge, L., Hagan, E., and Dhinojwala, A. (2008). Sticky gecko feet: The role of temperature and humidity. *Plos One*, e2192.
28. Persson, B. N. J. (2003). On the mechanism of adhesion in biological systems. *J. Chem. Phys.*, 7614–7621.
29. Persson, B. N. J., and Gorb, S. (2003). The effect of surface roughness on the adhesion of elastic plates with application to biological systems. *J. Chem. Phys.*, 11437–11444.
30. Pesika, N. S., Tian, Y., Zhao, B., Rosenberg, K., Zeng, H., McGuiggan, P., Autumn, K., and Israelachvili, J. N. (2007). Peel-zone model of tape peeling based on the gecko adhesive system. *J. Adhes.*, 383–401.
31. Puthoff, J. B., Holbrook, M., Wilkinson, M. J., Jin, K., Pesika, N. S., and Autumn, K. (2013). Dynamic friction in natural and synthetic gecko setal arrays. *Soft Matter*, 4855–4863.
32. Puthoff, J. B., Prowse, M. S., Wilkinson, M., and Autumn, K. (2010). Changes in materials properties explain the effects of humidity on gecko adhesion. *J. Exp. Biol.*, 3699–3704.

33. Rivlin, R. S. (1944). The effective work of adhesion. *Paint Technol.*, 215–218.

34. Ruibal, R., and Ernst, V. (1965). Structure of the digital setae of lizards. *J. Morphol.*, 271–293.

35. Russell, A. P. (1972). *The Foot of Gekkonid Lizards: A Study in Comparative and Functional Anatomy*. London, University of London.

36. Russell, A. P. (1975). A contribution to the functional morphology of the foot of the Tokay Gekko gecko. *J. Zool. London.*, 437–476.

37. Russell, A. P. (1976). Some comments concerning interrelationships amongst gekkonine geckos. *Morphology and Biology of Reptiles*. London, Academic Press, 217–244.

38. Russell, A. P. (1979). Parallelism and integrated design in the foot structure of gekkonine and diplodactyline geckos. *Copeia*, 1–21.

39. Russell, A. P. (1981). Descriptive and functional anatomy of the digital vascular system of the digital vascula. *J. Morphol.*, 293–323.

40. Russell, A. P. (1986). The morphological basis of weight-bearing in the scansors of the tokay gecko (Reptilia: Sauria). *Can. J. Zool.*, 948–955.

41. Russell, A. P. (2002). Integrative functional morphology of the gekkotan adhesive system (reptilia: gekkota). *Int. Comp. Biol.*, 1154–1163.

42. Russell, A. P., and Higham, T. E. (2009). A new angle on clinging in geckos: Incline, not substrate, triggers the deployment of the adhesive system. *Proc. R. Soc. B.*, 3705–3709.

43. Russell, A. P., and Johnson, M. K. (2007). Real-world challenges to, and capabilities of, the gekkotan adhesive system: Contrasting the rough and the smooth. *Can. J. Zool.*, 1228–1238.

44. Schmidt, H. R. (1904). Zur Anatomie und Physiologie der Geckopfote. *Jena Z. Naturalwiss.*

45. Stark, A. Y., Badge, I., Wucinich, N. A., Sullivan, T. W., Niewiarowski, P. H., and Dhinojwala, A. (2013). Surface wettability plays a significant role in gecko adhesion underwater. *Proc. Nat. Acad. Sci.*, 6340–6345.

46. Sun, W., Neuzil, P., Kustandi, T. S., O. H. S., and Samper, V. D. (2005). The nature of the gecko lizard adhesive force. *Biophys. J.*, 14–17.

47. Tian, Y., Pesika, N. S., Zeng, H., Rosenberg, K., Zhao, B., McGuiggan, P., Autumn, K., and Israelachvili, J. N. (2006). Adhesion and friction in gecko toe attachment and detachment. *Proc. Natl. Acad. Sci.*, 19320–19325.

48. Sun, W., Neuzil, P., Kustandi, T. S., Oh, S., and Samper, V. D. (2005). The nature of the gecko lizard adhesive force. *Biophys. J.*, L14–L17.
49. Williams, E. E., and Peterson, J. A. (1982). Convergent and alternative designs in the digital adhesive pads of scincid lizards. *Science.*, 1509–1511.

Chapter 5

From Sand Tube to Test Tube: The Adhesive Secretion from Sabellariid Tubeworms

Elise Hennebert,[a] Barbara Maldonado,[b] Cécile Van De Weerdt,[b] Mélanie Demeuldre,[b] Katharina Richter,[c] Klaus Rischka,[c] Patrick Flammang[a]

[a]*Biology of Marine Organisms and Biomimetics, University of Mons, 23 Place du Parc, 7000 Mons, Belgium*
[b]*Molecular Biology and Genetic Engineering Unit, GIGA, University of Liège, 1 Avenue de l'Hôpital, 4000 Liège, Belgium*
[c]*Department of Adhesive Bonding Technology and Surfaces, Fraunhofer Institute for Manufacturing Technology and Advanced Materials IFAM, 12 Wiener Strasse, 28359 Bremen, Germany*

Patrick.Flammang@umons.ac.be

Polychaetes of the family Sabellariidae are tube-dwelling marine worms commonly called honeycomb worms or sandcastle worms because they can build large colonies made up of thousands of tubes placed side by side. Tubes are composed of sand grains or shell fragments assembled with spots of cement released at the level of a bilobed building organ located near the mouth. This building organ is the visible external part of a complex secretory machinery made up of bouquets of cement cells located deep within the anterior part of the animal's body. Two main types of cement cells can been distinguished: cells with homogeneous granules and cells with

Bioadhesion and Biomimetics: From Nature to Applications
Edited by Havazelet Bianco-Peled and Maya Davidovich-Pinhas

ISBN 978-981-4463-98-0 (Hardcover), 978-981-4463-99-7 (eBook)
www.panstanford.com

heterogeneous granules containing inclusions. After secretion, the granules mix partially and form porous cement discs in which the pores derive from the swelling of the intragranular inclusions. The cement consists of several proteins, a sulphated polysaccharide, and magnesium and calcium ions. Five major proteins were identified which are characterized by highly repetitive and blocky primary structures. DOPA and phosphoserine, two post-translationally modified amino acids, were detected in some of the adhesive proteins. Biomimetic molecules have been produced in the form of recombinant preparations of the adhesive proteins, of chemically synthesized peptides, or by the functionalization of (meth)acrylate polymers with reactive DOPA groups. The bio-inspired polymers were condensed into a liquid complex coacervate exhibiting ideal features for effective underwater adhesion, allowing them to be tested in medical applications.

5.1 Introduction

Sabellariids are tube-dwelling marine polychaetes which live in the intertidal zone [1,2]. They are commonly called honeycomb worms or sandcastle worms because some species are gregarious and the tubes from a huge number of individuals are closely imbricated to form large reef-like mounds (Fig. 5.1a). To build the tube in which they live, they use their tentacles to collect particles such as sand grains, mollusc shell fragments or broken sea urchin spines from the water column and sea bottom, and convey them to the so-called building organ. This organ is a pulpy crescent-shaped bilobed structure located near the mouth (Fig. 5.1b). It selects the particles, dabs them with spots of cement, and adds them at the growing edge of the tube [3–5].

The stonemason abilities of sabellariids have been known for centuries [6], but it is only recently that some of the secrets of their cement have been unravelled, due to its potential to inspire the development of novel biomimetic underwater adhesives [see, e.g., Ref. 7]. As of today, the cement of sabellariid tubeworm is one of the best characterized marine bioadhesive together with those of mussels and barnacles [8,9]. Two species have attracted most of the attention in terms of tube building: *Sabellaria alveolata*, a common species along European coasts occurring from Cornwall to the south of Morocco [10]; and *Phragmatopoma californica*, which is found on the west coast of North America, from California to Mexico [11]. These species were investigated extensively in terms of building organ morphology, tube structure, and micro- and nanostructure as well as composition of the cement [3–5,12–18].

Figure 5.1 Part of a reef formed by the imbrication of the tubes built by the polychaete *Sabellaria alveolata* (a) and one individual of this species extracted from its tube (b). BO: building organ, PS: parathoracic segments, T: tentacule.

5.2 Characterization of the Natural Adhesive System

5.2.1 Morphology of the Adhesive Cells

The two lobes of the building organ are only the visible external part of a complex secretory organ made up of bouquets of cement cells located deep within the parathoracic segments of the worms

(Fig. 5.1b), mainly around the digestive tube (Fig. 5.2a) and at the basis of the parapodia [3,5].

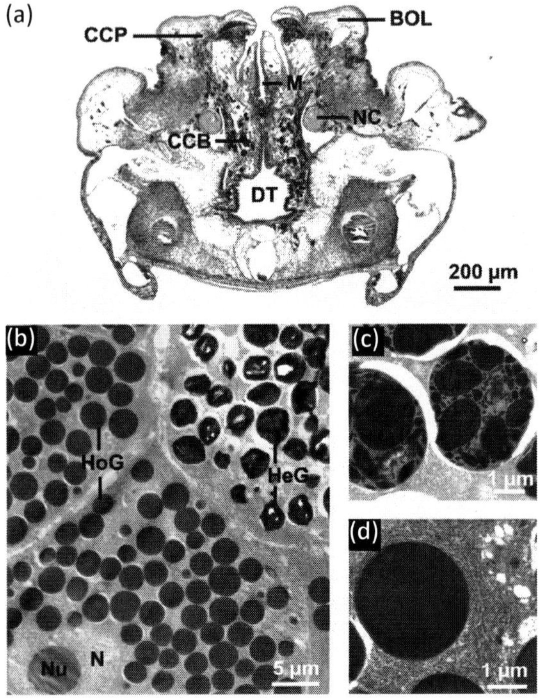

Figure 5.2 Morphology of the adhesive organ in *Sabellaria alveolata*. (a) Transverse section through the anterior part of the worm at the level of the mouth (light microscopy). (b) Ultrastructure of the two types of cement cells (TEM). (c and d) Detailed views of the heterogeneous and homogeneous granules, respectively (TEM). BOL: building organ lobe, CCB: cement cell bodies, CCP: cement cell processes, DT: digestive tube, HeG: heterogeneous granule, HoG; homogeneous granule, M: mouth, N: nucleus, NC: nerve cord, Nu: nucleolus.

These bouquets are groups of cement cell bodies that each emits one long cellular process (Fig. 5.2a). Both cell bodies and cell processes are filled with secretory granules. The cell processes are surrounded by loose connective tissue and interspersed with bands of smooth muscle tissue [3,5,19]. They meet at the level of the two ventral nerve cords before converging in bundles towards the tip of the lobes of the building organ (Fig 5.2a). When cement is applied as a small dab onto the mineral particles held in the building organ, secretory granules are delivered, apparently intact, at the surface of the building organ through a pair of pit-like openings (Fig. 5.2a) [3,5,17,19].

Two main types of cement cells have been distinguished based on the tinctorial affinities of their secretory granules [3,5]. Electron microscopy observations confirmed that two kinds of cement cells coexist: cells with homogeneous granules, accounting approximately for one third of the secretory cells, and cells with heterogeneous granules, accounting for the remaining two thirds (Fig. 5.2b) [5,18,19]. Both types of granules range from 2.5 to 4 μm in diameter. In homogeneous granules, no internal structure is visible (Fig. 5.2c), while heterogeneous granules enclose 100–1500 nm-large, spherical to elliptical inclusions (Fig. 5.2d). These inclusions are made up of electron-dense concentric lamellate layers and are embedded in a moderately electron-dense matrix resembling the content of homogeneous granules (Fig. 5.2c,d) [19].

Other cell types enclosing smaller granules (<1 μm) have been described in the building organ lobes but their involvement in cement production has not been investigated yet [5].

5.2.2 Morphology and Mechanics of the Cement

In the laboratory, the worms can build their tubes with different materials, including clean sand, bovine cortical bone fragments, diamond-shaped particles of silicon, or glass beads [4,14,20]. Because they are transparent, tubes made up of glass beads allow the direct observation of their organization (Fig. 5.3a). Beads are usually connected one to another by four to five cement discs with diameters ranging from 100 to 160 μm [4,14,18,19].

In scanning electron microscopy (SEM), cement discs display a smooth outer skin both at their margin and at the interface with the glass beads (Fig. 5.3b). When their inner core is revealed by cohesive fracture, however, it appears as porous [4,19,20]. This porous structure is confirmed by observation of native cement dabs in confocal laser scanning microscopy (Fig. 5.3c) [14]. The pores, whose diameters range from 0.25 to 6.5 μm in *P. californica*, from 0.1 to 5 μm in *Phragmatopoma lapidosa*, and from 0.3 to 4 μm in *S. alveolata*, form a porosity gradient, the smallest pores being at the outside edges of the disc and the biggest ones at the centre [4,14,19,20]. In *S. alveolata*, the porous organisation of the cement was also observed in transmission electron microscopy (Fig. 5.3d)(TEM). Using this technique, it was shown that pores are actually hollow inclusions embedded in a matrix [19]. The cortex of these inclusions consists of an electron-dense lamellar material disposed in concentric layers. Electron-dense granules measuring

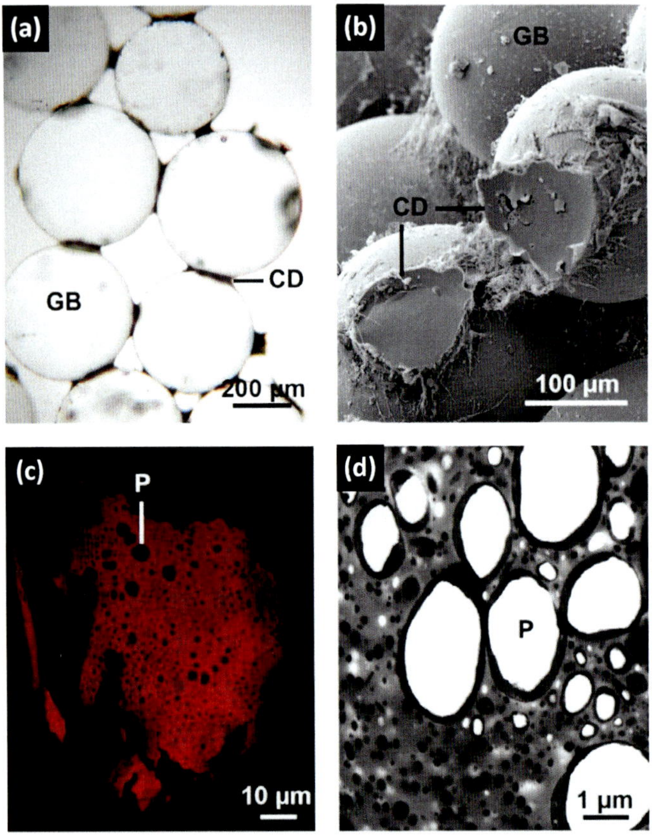

Figure 5.3 Morphology of the cement produced by *Sabellaria alveolata*. (a) Glass beads joined together by cement discs (light microscopy). (b) Outer aspect of the cement discs (scanning electron microscopy). (c) Optical section through an unstained cement disc (confocal laser scanning microscopy). The cement is autofluorescent across the visible spectrum [4]. (d) Ultrastructure of the cement showing the electron-dense material forming the cortex of the pores (transmission electron microscopy). CD: cement disc, GB: glass bead, P: pore.

50–700 nm in diameter and lacunae ranging from 50 to 2500 nm in diameter were also observed in the matrix [19]. These results indicate that the homogeneous material from the granules of the two types of cement cells forms the matrix of the cement while the pores derive directly from the inclusions of the heterogeneous granules which swell through an unknown process [3,19]. In *P. californica*, scanning electron microscopy observations using back-scattered electrons (SEM BSE) confirmed this structure and origin for the pores in the cement [18]. At the nanoscale, the cement in this species was found to be composed of globular structures measuring 50–100 nm in diameter using atomic force microscopy (AFM) [14].

Mechanical tests performed on the tubes of *S. alveolata* revealed that the cement behaves like a visco-elastic material [21]. This property allows the tubes to dissipate the mechanical energy from the waves to which they are subjected and to reduce the mechanical stress transmitted inside the tubes to the worms. The visco-elastic behaviour of the cement appears to be maintained for years [21]. In *P. californica*, Sun and co-workers measured the pull-out or detachment force of individual sand grains from the tubes [22]. An approximate adhesive strength was then calculated from the mean sand grain pull-out force of 25 mN. Given that each sand grain or glass bead is typically held in place by four to five spots of cement, an adhesive strength of 350 kPa can be estimated for one cement disc of 150 μm in diameter.

5.2.3 Cement Composition and Formation

Cement composition has been extensively investigated in the species *P. californica*, and consists mostly of several different proteins, a sulphated polysaccharide, and large amounts of Mg^{2+} and Ca^{2+} ions [13,14,18,23]. Five major proteins, named *Phragmatopoma* cement proteins (abbreviated as Pc-1 to -5) and a tyrosinase-like enzyme have been identified and presented significant expression levels in the worm parathoracic region [17,18]. The five cement proteins have highly repetitive and blocky primary structures with limited amino acid diversity (Table 5.1; [13,17,18]). Pc-1 (18 kDa) is a glycine and tyrosine-rich protein displaying a 10-amino acid repeated motif and a basic isoelectric point (pI) due to the abundance of lysine residues. Pc-2, Pc-4 and Pc-5 (20, 27 and 15 kDa, respectively) are all basic, histidine-rich proteins with repeated motifs. Pc-3 occurs in the form of two variants (Pc-3A and Pc-3B) both characterized by the over-abundance of serine residues which account for 60–90 mol% of the protein. Most of these serine residues are phosphorylated and Pc-3 is therefore extremely acidic with a calculated pI ranging from 0.5 to 1.5 [4,13]. Noteworthily, tyrosine residues of both Pc-1 and Pc-2 are post-translationally modified into 3,4-dihydroxyphenylalanine (DOPA) by the addition of a hydroxyl group [24]. Some of these DOPA residues are further halogenated into 2-chloro-DOPA residues [25]. In *S. alveolata*, the primary sequences of the proteins Sa-1, -2 and -3 were obtained directly from cDNAs amplified using primers designed from their homologues Pc-1, -2 and -3, therefore allowing to bypass the extraction, purification and characterization of the proteins [16]. In this species, DOPA and phosphoserine (pSer) residues were highlighted through histochemical and immunohistochemical detection [3,16,26] and, by homology, it is likely that DOPA residues would occur in Sa-1 and -2 whereas pSer residues would be present only in Sa-3. Although the American and European species share related adhesive proteins with similar characteristics, alignment of the amino acid sequences of the proteins from both species resulted in limited percentages of identity [16]. Therefore, the main features of the adhesive proteins such as amino acid composition, post-translational modifications, presence of repeated patterns, pI and mass, appear to be more important for the function than the primary structure.

Table 5.1 Characteristics of the cement proteins from *Phragmatopoma californica* and *Sabellaria alveolata*

Protein	Mass (kDa)	Consensus repeated unit (frequency)	Features	Ref.
Pc1	18	VGGYGYGGKK (15)	DOPA, glycine- and tyrosine-rich, basic	[13]
Pc2	21	HPAVXHKALGGYG(8)	DOPA, histidine-rich, basic	[13]
Pc3	10–52[a]	Poly(S)	p-Ser (about 80 mol%), highly acidic	[13]
Pc4	14	GVGLHGAGYGLHXGVGLHG (5)	histidine-rich,basic	[23]
Pc5	24	HGGWHAP (3)	histidine-rich, basic	[23]

Sa1	22	KGAYGAKGLGYGNKAGYGAYG (6–8)	glycine- and tyrosine-rich, basic, DOPA[b]	[16]
Sa2	22	VHKAAWG (5) and VHKAAGYGGYG (8)	histidine-rich, basic, DOPA[b]	[16]
Sa3	(21–22)[a]	Poly(S)	p-Ser[b], highly acidic	[16]

[a]The masses of Pc3 and Sa3 vary according to the variant considered.
[b]From indirect evidence.

In *P. californica*, the co-occurrence of positively and negatively charged proteins in the cement led to the hypothesis that a phenomenon called complex coacervation could play a role in the condensation of the adhesive [4,13]. Coacervation occurs when the charges of the polyelectrolytes are balanced, leading to the formation of a dense, solute-rich, water-immiscible fluid. Recently, however, Stewart et al. [7] and Wang and Stewart [17] pointed out that the rapidity of cement setting after release (i.e. about 30 s; [14]) and the poor mixing of the oppositely charged proteins within the cement are not consistent with the process of complex coacervation at the whole cement level. Nevertheless, in cement cells, at the granule level, electrostatic charge neutralization between oppositely charged polyions could drive the condensation of adhesive proteins [17,18]. Indeed, histochemistry, immunohistochemistry and *in situ* hybridization experiments localized Pc-1, Pc-3A/B and Pc-4 in the cement cells containing heterogeneous granules, and Pc2, Pc-5 and the sulphated polysaccharide in the cement cells with homogeneous granules [17,18]. As for the tyrosinase-like enzyme, it was expressed in both cement cell types [17]. In both *P. californica* and *S. alveolata*, energy dispersive x-ray spectroscopic (EDS) analyses highlighted the presence of P (presumably from Pc-3A/B) and Mg in the inclusions of the heterogeneous granules [7,12]. All these results taken together allowed Stewart and co-authors to propose a model for cement production in *P. californica* [18], model that could be extended to *S. alveolata* [16]. The homogeneous granules, based on their morphology, contain a uniformly distributed complex of polybasic proteins (Pc-2 and Pc-5) and sulphated polysaccharides. The organization of the heterogeneous granules is more complicated, the polyphosphates (Pc-3A/B) being colocated with Mg^{2+} exclusively in the inclusions, and surrounded by a uniform matrix containing the polybasic proteins (Pc-1 and Pc-4). After secretion, the two types of complex coacervates would fuse together to form the final cement and, although individual granules are no longer visible in the final structure, their contents appear as poorly mixed [17,18].

Among the functional groups brought by the different cement components, the modified amino acids present in the different adhesive proteins (i.e. DOPA and pSer) and the sulphate groups borne by the polysaccharide would mediate physicochemical interactions with the substratum (adhesion) and would also be involved in cement solidification (cohesion) [4,7,13,17,18,25]. Phosphates and sulphates may be involved in non-covalent adhesive and/or cohesive interactions, possibly through Ca^{++} or Mg^{++} bridging [7,22,26]. Indeed, chelation of these divalent cations by EDTA lowered both the maximal adhesive force and energy dissipation of cement by up to an order of magnitude [22]. DOPA groups take part in surface coupling either through hydrogen bonds or by forming complexes with metal ions and metal oxides present in mineral surfaces [27]. Following oxidation, DOPA groups also contribute to cement curing by forming intermolecular cross-links (e.g. cysteinyl-DOPA cross-links; [13]). This oxidation is presumably catalysed by the tyrosinase-like enzyme co-secreted with the Pcs [17,18].

5.3 Bio-Inspired Adhesive Polymers: Production, Characterization and Applications

To understand in details the role of each of the components of the cement, especially the adhesive proteins, and the way they auto-assemble to form a strong adhesive, it is necessary to obtain these components in large quantities. To bypass the problem of extracting adhesive proteins directly from worms, biomimetic molecules have been produced either in the form of recombinant preparations of the adhesive proteins or of chemically synthesized peptides. Although these two approaches have been used extensively for the production of mussel-mimetic molecules [9,27], only preliminary studies have been conducted for tubeworm adhesive proteins.

Another way to mimic adhesive proteins is by the chemical synthesis of important functional parts of these adhesives. This has been done by the functionalization of various polymers with reactive DOPA or catechol groups. These polymers have then been combined to develop bio-inspired adhesives mostly for biomedical applications [7,28,29].

5.3.1 Recombinant Proteins

Recombinant DNA technology has been used to produce one of the adhesive proteins of *S. alveolata*, the protein Sa-1 (Lejeune and Van de Weerdt, unpubl.). A codon-optimized Sa-1 gene was chemically synthesized and inserted in expression vectors (pET type) compatible with protein production in the bacterium *Escherichia coli* and containing a histidine tag sequence to allow purification of the protein using metal affinity chromatography (Fig 5.4a). The vector was successfully expressed in *E. coli* and the recombinant protein, rSa-1, was mainly produced in the insoluble fraction of the bacteria, probably as inclusion bodies. It was therefore extracted with a buffer containing strong chaotropic agents and was purified by chromatography on a column packed with a nickel-charged resin, under denaturing conditions. Yields of 50 mg of pure protein per litre of culture were obtained for rSa-1 (Fig. 5.4b).

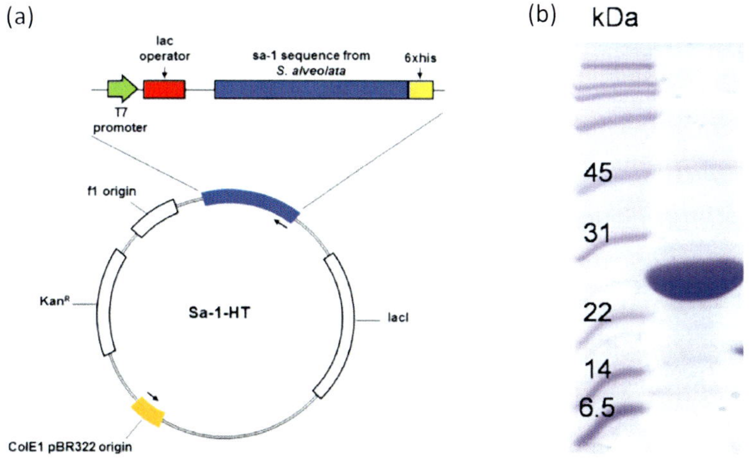

Figure 5.4 (a) Plasmid vector construction for the over-expression of recombinant Sa-1 in the bacterium *Escherichia coli*. (b) SDS-PAGE analysis after purification of recombinant Sa-1. Left lane: protein molecular weight markers; right lane: purified recombinant Sa-1.

There is, however, a drawback with the heterologous expression of Sa-1: the fact that the recombinant proteins lack the post-translational modifications such as the hydroxylation of tyrosine residues. The production of functional recombinant adhesive proteins therefore requires an additional enzyme-catalysed in vitro modification step of tyrosine residues into DOPA. This is usually done using a commercially available mushroom tyrosinase [28].

5.3.2 Synthetic Peptides

Chemical synthesis allows the direct incorporation of DOPA into biomimetic peptides, thus avoiding the complications associated with the oxidation of precursor tyrosine residues. A 21-amino acid peptide based on the consensus repeat sequence of Sa-1 (Table 5.1) and containing DOPA residues was prepared by solid phase peptide synthesis using Fmoc-chemistry (Rischka, unpubl.). Considering that all the tyrosine residues are hydroxylated, the amino acid sequence of the Sa-1 derived peptide is as follows:

H-Lys-Gly-Ala-DOPA-Gly-Ala-Lys-Gly-Leu-Gly-DOPA-Gly-Asn-Lys-Ala-Gly-DOPA-Gly-Ala-DOPA-Gly-OH

Due to the high amount of hydrophobic amino acids in the sequence, a high potential for the formation of secondary structures was expected during the synthesis, which would result in poor coupling yields. To circumvent this problem, Fmoc-(N-Tmob)Glycine was introduced instead of Fmoc-Glycine at position 8 (...Lys-Gly-Leu...) in the synthesis. The bulky Tmob-group (S-2,4,6-trimethoxybenzyl) is responsible for the disruption of secondary structures [30].

Quartz crystal microgravimetry was used to explore the adsorption kinetic of the obtained Sa-1 peptide on silica and titanium coated sensors (Fig. 5.5; Richter and Rischka, unpubl.). The adsorption properties of the Sa-1 peptide were compared with another DOPA containing peptide known to adhere well to these surfaces, the Mefp-1 peptide from the blue mussel [31]. The adsorption of Sa-1 and Mefp1 on titanium showed similar kinetics (Fig. 5.5). In accordance with the Rollover Model for protein adsorption [32], a reorientation from end-on adsorbed peptides to side-on peptides might cause the overshooting kinetics. On the contrary, the adsorption on silica showed a kinetic corresponding

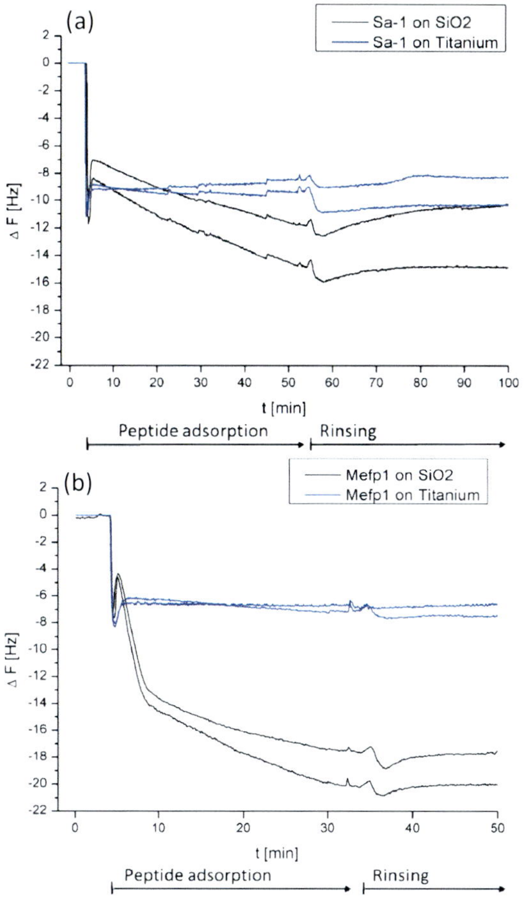

Figure 5.5 QCM-D measurements of the adsorption of Sa-1 (a) and Mefp-1 (b) peptides on titanium and silica surfaces. Before each measurement, the peptides Sa-1 (KGA**Y**GAKGLG**Y**GNKAG**Y**GA**Y**G) and Mefp-1 (AKPSY**PP**T**Y**K) were freshly dissolved in dd-H_2O at a concentration of 1 mg/mL. The measurements were run under flow conditions of 100 μL/min at 25°C. After a baseline against dd-H_2O had established, the peptide solution was flowed over the sensor. Then, the peptide solutions were substituted by dd-H_2O after 30 min (Mefp-1) and 50 min (Sa-1) and the sensor was rinsed in order to study the desorption behaviour of the adsorbed peptide. Measurements were performed on a Q-sense E4 device (Gothenburg, Sweden). **Y**: 3,4-Dihydroxyphenylalanine (DOPA), **P**: Hydroxyproline.

to the three-state model for proteins [33]. Assuming this model is applicable to peptides as well, three different states of adsorption follow one another: an irreversible initial state, a reversible intermediate state, and a final irreversible state. This might be the reason for more peptide adsorbed on silica compared to titanium. For both substrates, rinsing did not cause significant desorption. This could be explained by the strong interaction of DOPA with the oxygen atoms exposed by titanium (titanium forms a titanium oxide passivation layer) and silica [34–36].

5.3.3 Polymers

DOPA being considered as a key component for adhesion, functionalization of synthetic polymers other than polypeptides with catechol groups has emerged as a promising strategy for the development of new biomimetic adhesives [27–29]. Based on the *P. californica* model, poly(meth)acrylate polymers with phosphate, amine and catechol sidechains in the same molar ratios as the natural adhesive proteins Pc-1 and Pc-3 were synthesized [37]. Mimics of Pc-3 were synthesized by free radical copolymerization of monoacryloxyethyl phosphate (MAEP), acrylamide and dopamine methacrylate (DMA), the latter being a DOPA analogue monomer (Fig. 5.6a). Analogues of Pc-1 were created by copolymerization of N-(3-aminopropyl)methacrylamide hydrochloride (APMA) and acrylamide (Fig. 5.6b). Both copolymers were mixed, with divalent cations, in similar proportions as the natural proteins. Around neutral pH, they condensed into a liquid complex coacervate exhibiting several features, including phase separation from water and low interfacial tension, which are invaluable for effective underwater adhesion [37]. The underwater bond strength of the biomimetic adhesive on aluminium was ~0.8 MPa, about twice the value estimated for the natural glue [37]. To increase bond strengths, a second polymer network can be created within cross-linked coacervate network by entrapping polyethylene glycol diacrylate (PEG-dA) monomers in the coacervate phase. Simultaneous polymerization of PEG-dA and cross-linking of the coacervate network resulted in maximum shear bond strengths of ~1.2 MPa [38]. The bond strength and complex fluid behaviour of the polymer-based coacervates have important implications for medical applications of the adhesives.

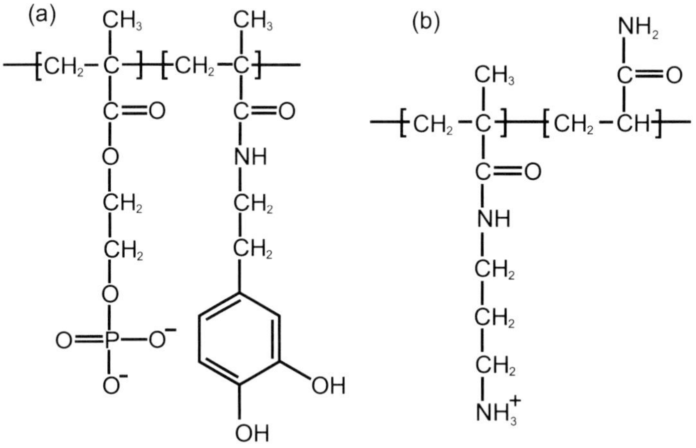

Figure 5.6 Synthetic analogs of cement proteins from *Phragmatopoma californica*. (a) Structure of the Pc-3 polymethacrylate analog copolymer. (b) Structure of the Pc-1 polyamine analog copolymer. The bio-inspired polymers are random copolymers synthesized by free radical polymerization [37].

The PEG-dA-filled adhesive coacervate, used in conjunction with a patch, was tested for sealing foetal membrane defects. The sealant withstood mechanical stress without leakage of fluid or slippage and was shown to be non-toxic when in direct contact with human foetal membranes in an organ culture setting [39]. A biodegradable, temperature-triggered version of the adhesive coacervate was also designed by replacing the APMA/acrylamide copolymer by an amine-modified collagen hydrolysate [40]. This adhesive coacervate was used to repair rat calvarial bone defect and was capable of maintaining three-dimensional bone alignment in freely moving rats over a 12-week indwelling period [41]. Histological evaluation demonstrated that the adhesive was gradually resorbed and replaced by new bone with an inflammatory response commensurate with normal wound healing. This non-cytotoxic degradable coacervate therefore appears to be suitable for use in the reconstruction of craniofacial features [41].

Acknowledgements

This work was supported by the "Service Public de Wallonie—Programme Winnomat 2", by the "Communauté française de

Belgique—Actions de Recherche Concertées" and by COST Action TD0906 (http://www.cost-bioadhesives.org/). EH and BM, and PF are, respectively, postdoctoral researchers and research director of the Fund for Scientific Research of Belgium (FNRS). This study is a contribution of the Centre Interuniversitaire de Biologie Marine (CIBIM).

References

1. Rouse G., and Pleijel F. (2001). *Polychaetes*. Oxford University Press, Oxford, 354p.
2. Capa M., Hutchings P., and Peart R. (2012). Systematic revision of Sabellariidae (Polychaeta) and their relationships with other polychaetes using morphological and DNA sequence data. *Zool. J. Linnean Soc.*, **164**, 245–284.
3. Vovelle J. (1965). Le tube de *Sabellaria alveolata* (L.): Annélide polychète Hermellidae et son ciment. Etude écologique, expérimentale, histologique et histochimique. *Arch. Zool. Exp. Gen.*, **106**, 1–187.
4. Stewart R. J., Weaver J. C., Morse D. E., and Waite J. H. (2004). The tube cement of *Phragmatopma californica*: A solid foam. *J. Exp. Biol.*, **207**, 4727–4734.
5. Wang C. S., Svendsen K. K., and Stewart R. J. (2010). Morphology of the adhesive system in the sandcastle worm, *Phragmatopoma californica*. In (von Byern J. and Grunwald I., eds.), *Biological Adhesive Systems—From Nature to Technical and Medical Application*, Springer, Wien, pp. 169–179.
6. Réaumur R. A. Ferchault de. (1711). Des différentes manières dont plusieurs espèces d'animaux de mer s'attachent au sable, aux pierres et les uns aux autres. *Mém. Acad. R. Sci.*, 108–134.
7. Stewart R. J., Wang C. S., and Shao H. (2011). Complex coacervates as a foundation for synthetic underwater adhesives. *Adv. Colloid Interface Sci.*, **167**, 85–93.
8. Kamino K. (2010). Molecular design of banacle cement in comparison with those of mussel and tubeworm. *J. Adhes.*, **86**, 96–110.
9. Hennebert E., Becker P., and Flammang P. (2012). Lessons from sea organisms to produce new biomedical adhesives. In (Mano J. F., ed.), *Biomimetic Approaches for Biomaterial Development*. Wiley-VCH, Weinheim, pp. 273–291.

10. Dubois S., Barillé L., and Retière C. (2003). Efficiency of particle retention and clearance rate in the polychaete *Sabellaria alveolata* L. C. R. *Biologies*, **326**, 413–421.
11. Drake C. A., McCarthy D. A., and von Dohlen C. D. (2007). Molecular relationships and species divergence among *Phragmatopoma* spp. (Polychaeta: Sabellaridae) in the Americas. *Mar. Biol.*, **150**, 345–358.
12. Gruet Y., Vovelle J., and Grasset M. (1987). Composante biominérale du ciment du tube chez *Sabellaria alveolata* (L.), annélide polychète. *Can. J. Zool.*, **65**, 837–842.
13. Zhao H., Sun C., Stewart R. J., and Waite J. H. (2005). Cement proteins of the tube-building polychaete *Phragmatopoma californica*. *J. Biol. Chem.*, **280**, 42938–42944.
14. Stevens M. J., Steren R. E., Vlamidir H., and Stewart R. J. (2007). Multiscale structure of the underwater adhesive of *Phragmatopoma californica*: A nanostructured latex with a steep microporosity gradient. *Langmuir*, **20**, 5045–5049.
15. Fournier J., Etienne S., and Le Cam J.-B. (2010). Inter- and intraspecific variability in the chemical composition of the mineral phase of cements from several tube-building polychaetes. *Geobios*, **43**, 191–200.
16. Becker P. T., Lambert A., Lejeune A., Lanterbecq D., and Flammang P. (2012). Identification, characterization, and expression levels of putative adhesive proteins from the tube-dwelling polychaete *Sabellaria alveolata*. *Biol. Bull.*, **223**, 217–225.
17. Wang C. S. and Stewart R. J. (2012). Localization of the bioadhesive precursors of the sandcastle worm, *Phragmatopoma californica* (Fewkes). *J. Exp. Biol.*, **215**, 351–361.
18. Wang C. S. and Stewart R. J. (2013). Multipart copolyelectrolyte adhesive of the sandcastle worm, *Phragmatopoma californica* (Fewkes): Catechol oxidase catalyzed curing through peptidyl-Dopa. *Biomacromolecules*, **14**, 1607–1617.
19. Flammang P. and Lambert A. (2008). Cement ultrastructure and adhesive glands morphology in the tube-dwelling polychaete *Sabellaria alveolata*. *J. Morphol.*, **269**, 1479.
20. Dean M., Welch J., Brandt C., and Tauer T. (2009). Surface analyses of biocements from *Pectinaria gouldii* (Polychaeta: Pectinariidae) and *Phragmatopoma lapidosa* (Polychaeta: Sabellariidae). *Zoosymposia*, **2**, 329–337.

21. Le Cam J.-B., Fournier J., Etienne S., and Couden J. (2011). The strength of biogenic sand reefs: Visco-elastic behaviour of cement secreted by the tube building polychaete *Sabellaria alveolata*, Linnaeus, 1767. *Estuar. Coast. Shelf Sci.*, **91**, 333–339.
22. Sun C., Fantner G. E., Adams J., Hansma P. K., and Waite J. H. (2007). The role of calcium and magnesium in the concrete tubes of the sandcastle worm. *J. Exp. Biol.*, **210**, 1481–1488.
23. Endrizzi B. J. and R. J. Stewart. (2009). Glueomics: An expression survey of the adhesive gland of the sandcastle worm. *J. Adhes.*, **85**, 546–559.
24. Waite J. H., Jensen R. A., and Morse D. E. (1992). Cement precursor proteins of the reef-building polychaete *Phragmatopoma californica* (Fewkes). *Biochemistry*, **31**, 5733–5738.
25. Sun C., Shrivastava A., Reifert J. R., and Waite J. H. (2009). Halogenated DOPA in a marine adhesive protein. *J. Adhes.*, **85**, 126–138.
26. Flammang P., Lambert A., Bailly P., and Hennebert E. (2009). Polyphosphoprotein-containing marine adhesives. *J. Adhes.*, **85**, 447–464.
27. Lee, B. P., Messersmith P. B., Israelachvili J. N., and Waite J. H. (2011). Mussel-inspired adhesives and coatings. *Annu. Rev. Mater. Res.*, **41**, 99–132.
28. Stewart R. J. (2011). Protein-based underwater adhesives and the prospects for their biotechnological production. *Appl. Microbiol. Biotechnol.*, **89**, 27–33.
29. Brubaker C. E. and Messersmith P. B. (2012). The present and future of biologically inspired adhesive interfaces and materials. *Langmuir*, **28**, 2200–2205.
30. Johnson T., Quibell M., and Sheppard R. C. (1995). N,O-bisFmoc derivatives of N-(2-hydroxy-4-methoxybenzyl)-amino acids: Useful intermediates in peptide synthesis. *J. Peptide Sci.*, **1**, 11–25.
31. Richter K., Diaconu G., Rischka K., Amkreutz M., Müller F. A., and Hartwig A. (2013). Adsorption studies of mussel-inspired peptides. Bioinspired, *Biomimetic Nanobiomater.*, **2**, 45–53.
32. Wertz C. and Santore M. (2002). Adsorption and reorientation of lysozyme on hydrophobic surfaces. *Langmuir*, **18**, 1190–1199.
33. Rabe M., Verdes D., and Seeger S. (2011). Understanding protein adsorption phenomena at solid surfaces. *Adv. Colloid Interface Sci.*, **162**, 87–106.

34. Li S. C., Chu L. N., Gong X. Q., and Diebold U. (2010). Hydrogen bonding controls the dynamics of catechol adsorbed on a TiO_2 (110) surface. *Science*, **328**, 882–884.

35. Qin Z. and Buehler M. (2012). Molecular mechanics of dihydroxyphenylalanine at a silica interface. *Appl. Phys. Lett.*, **101**, 083702.

36. Yu J., Wei W., Menyo M. S., Masic A., Waite J. H., and Israelachvili J. N. (2013). Adhesion of mussel foot protein-3 to TiO_2 surfaces: the effect of pH. *Biomacromolecules*, **14**, 1072–1077.

37. Shao H., Bachus K. N., and Stewart R. J. (2008). A water-borne adhesive modeled after the sandcastle glue of *P. californica*. *Macromol. Biosci.*, **9**, 464–471.

38. Kaur S., Mahika Weerasekare G., and Stewart R. J. (2011). Multiphase adhesive coacervates inspired by the sandcastle worm. *ACS Appl. Mater. Interfaces*, **3**, 941–944.

39. Mann L. K., Papanna R., Moise Jr. K. J., Byrd R. H., Popek E. J., Kaur S., Tseng S. C. G., and Stewart R. J. (2012). Fetal membrane patch and biomimetic adhesive coacervates as a sealant for fetoscopic defects. *Acta Biomater.*, **8**, 2160–2165.

40. Shao H. and Stewart R. J. (2010). Biomimetic underwater adhesives with environmentally triggered setting mechanisms. *Adv. Mater.*, **22**, 729–733.

41. Winslow B. D., Shao H., Stewart R. J., and Tresco P. A. (2010). Biocompatibility of adhesive complex coacervates modelled after the sandcastle glue of *Phragmatopoma californica* for craniofacial reconstruction. *Biomaterials*, **31**, 9373–9381.

Section III

Biomimetic Adhesives

Chapter 6

Adhesives and Coatings Inspired by Mussel Adhesive Proteins

Hao Meng,[a] Yuan Liu,[a] Morgan M. Cencer,[b] and Bruce P. Lee[a]

[a]*Department of Biomedical Engineering and*
[b]*Department of Chemistry, Michigan Technological University,*
1400 Townsend Dr. Houghton, MI 49931, USA
bplee@mtu.edu

Mollusks such as the blue mussel (*Mytilus edulis*) secrete adhesive proteins that exhibit strong and reliable underwater adhesion. A key adhesive component in these adhesive proteins is an amino acid, 3,4-dihydroxyphenylalanine (DOPA), which is responsible for both interfacial binding and intermolecular cross-linking. DOPA is a unique and versatile adhesive moiety, capable of binding to both inorganic and organic surfaces through either strong reversible bonds or covalent attachment. This chapter reviews the chemistry of DOPA side chain and the use of DOPA and its derivatives (e.g., dopamine) as building blocks in developing mussel-inspired adhesives, coatings, and multifunctional polymeric anchors for various applications.

6.1 Introduction

Performance of most man-made adhesives is significantly compromised in the presence of moisture, which has traditionally

Bioadhesion and Biomimetics: From Nature to Applications
Edited by Havazelet Bianco-Peled and Maya Davidovich-Pinhas

ISBN 978-981-4463-98-0 (Hardcover), 978-981-4463-99-7 (eBook)
www.panstanford.com

been treated as a surface contaminant that needs to be removed to prevent the formation of a weak boundary layer [1]. In many applications, and particularly those in the medical field, development of moisture-resistant adhesives is in high demand. Marine mussel, a bivalve mollusk, secretes protein-based adhesives to secure itself to wet surfaces in the turbulent intertidal zone [2]. 3,4-dihydroxyphenylalanine (DOPA, Fig. 6.1a), an amino acid post-translationally modified from tyrosine, is found in large abundance in these adhesive proteins and is responsible for both curing of the adhesive and interfacial binding [3–5]. DOPA is a unique and versatile adhesive molecule, capable of binding to both inorganic and organic surfaces through either strong reversible bonds or covalent attachment [6,7].

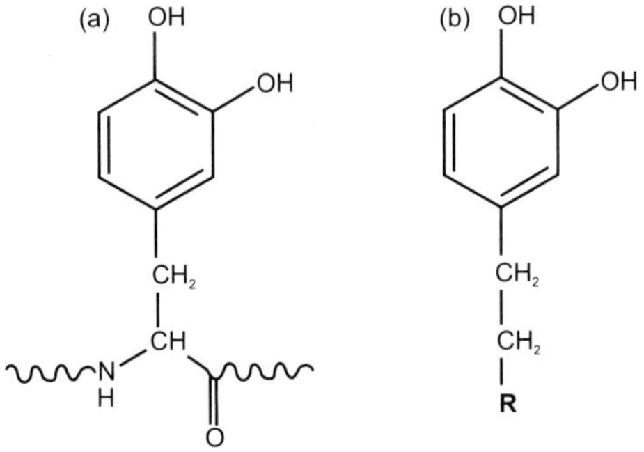

Figure 6.1 Chemical structure of DOPA (a) and other catechol derivatives (R = $-NH_2$ and $-CO_2H$ for dopamine and 3,4-dihydroxyhydrocinnamic acid, respectively) (b) used in the preparation of synthetic mussel-mimetic adhesive polymers.

Mussel adhesive proteins (MAPs) extracted from mollusks such as the common blue mussel (*Mytilus edulis*) have yielded strong adhesion [8,9]. However, several thousand specimens of mussels are needed for 1 gram of protein [10], which makes it highly impractical for adhesive production through this technique. Recent advances have focused on the mechanistic understanding of DOPA chemistry, in the hope of developing DOPA-containing synthetic adhesives. Modification of DOPA and its derivatives, such as dopamine and 3,4-dihydroxyhydrocinnamic acid (Fig. 6.1b), onto inert, synthetic polymers has imparted these materials with strong, water-resistant adhesive properties to various biological substrates including soft tissue [11–13], mucin [14], bone [15], and various metallic [16–22] and polymeric [23–26] substrates. This chapter reviews the chemistry of the catechol side chain of DOPA and the utilization of this versatile building block for designing mussel-inspired adhesives and coatings. Finally, the ability for dopamine to auto-polymerize and function as a facile polymeric anchor is introduced.

6.2 Chemistry of DOPA Side Chain

DOPA is capable of undergoing several types of catechol–catechol and catechol–surface interactions, leading to the curing of the catechol-containing adhesive and strong interfacial binding (Fig. 6.2). Additionally, catechol is a unique molecule capable of forming strong bonds to both inorganic and organic substrates using either reversible physical or irreversible covalent cross-links.

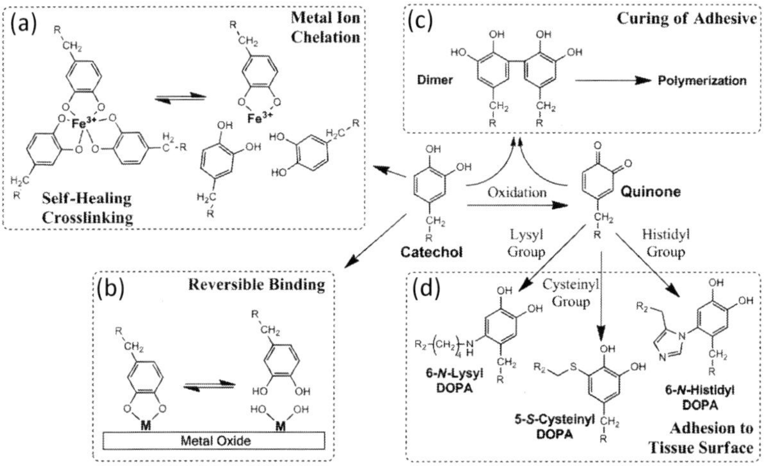

Figure 6.2 Catechol side chain of DOPA is capable of various catechol–catechol and catechol–surface interactions through either reversible physical or irreversible covalent interactions. Catechol chelates transitional metal ions, which leads to self-healing cross-linking (a). Catechol forms strong reversible coordination bonds with metal oxide surfaces (b). Oxidation of catechol forms highly reactive quinone, which results in covalent cross-linking with other catechol moieties and polymerization (c). Quinone reacts with various functional groups (i.e., $-NH_2$, $-SH$) found on biological substrates resulting in interfacial covalent cross-linking (d).

6.2.1 Reversible Interactions

Mussels are known to accumulate metal ions in their byssus threads at a concentration significantly higher than that of the surrounding sea water due to a high catechol content [27,28]. Catechol is capable of forming strong complexes with metal ions (Fe^{3+}, Ca^{2+}, Cu^{2+}, Ti^{3+}, Ti^{4+}, Mn^{2+}, Mn^{3+}, Zn^{2+}) with log stability constants greater than 40 (at 20°C, pH 10, log stability constant = 47 and 45 for Al^{3+} and Fe^{3+}, respectively) [2,29–31]. Taylor et al. [31,32] have shown that synthetic peptides that mimic MAPs form bis- and tris-catecholate complexes with Fe^{3+} depending on the ratio of DOPA to ferric ion (Fig. 6.2a). Breaking and reformation of strong, reversible catechol–metal ion interaction contributes to the wear resistance properties of the protective coatings on mussel byssus threads, which experience large cyclic strains in the turbulent intertidal zone [33,34]. This interaction was incorporated into synthetic hydrogels to form self-healing networks with elastic moduli approaching those of covalently cross-linked hydrogels [35].

The affinity of catechols for metals is not limited to soluble metal ions but is also extended to metal oxide surfaces (Fig. 6.2b) [36–39]. Catechol forms a strong, reversible bond with a titanium surface with a bond strength averaging around 800 pN, reaching 40% that of a covalent bond (2000 pN for carbon–silicon bond) [40]. This is the strongest reversible bond involving a biological molecule reported to date (Table 6.1). Density functional theory analysis revealed that catechol readily displaces water molecules and adsorbs directly to the metal oxide surface with binding energy reaching 22–33 kcal/mol [41–43]. The ability to adsorb to various metal oxide surfaces makes DOPA a versatile anchoring group for surface modification of these substrates. Dalsin et al. [16–18] have reported the use of PEGylated DOPA and DOPA-containing peptides to render metal substrates (Au_2O_3, Al_2O_3, SiO_2 TiO_2, NiTi, and stainless steel) non-fouling to cell and protein adhesion.

The catechol also forms hydrogen bonds, which allow it to compete with water for hydrogen bonding sites and absorb onto mucosal tissues [14,44,45] and hydroxyapatite [46]. Finally, catechol residues can interact with other aromatic rings through π–π electron interaction, which may contribute to adhesion to surfaces rich in aromatic compounds such as polystyrene [47] and gold substrates [48].

Table 6.1 Single-molecular binding strength of biomolecules

Type of interaction	Binding strength (pN)	Reversible	Reference
Integrin-fibrinogen	80–90	Yes	[49]
Histidine-Ni	150–190	Yes	[50]
Avidin-biotin	160–200	Yes	[51,52]
Tyrosine-titanium	100	Yes	[40]
DOPA-titanium	**800**	**Yes**	**[40]**
Silicon-carbon	2000	No	[53]

6.2.2 Covalent Bond Formation

Catechol can be easily oxidized to highly reactive quinone, which is capable of forming covalent cross-links. Several types of enzymes (tyrosinase (EC 1.14.18.1), catechol oxidase (EC 1.10.3.1), laccase (EC 1.10.13.2), and peroxidase (EC 1.11.1.17)) catalyze the oxidation of catechols to form the corresponding quinones [54–56]. Chemical oxidation of catechols can also be achieved through utilizing oxidizing reagents such as hydrogen peroxide, periodate, and iron (III) [57–59]. Additionally, catechol undergoes auto-oxidation in aerated and basic aqueous solution [3,60,61].

Quinone is highly reactive and can participate in intermolecular covalent cross-linking, leading to the rapid hardening of catechol-containing adhesives and interfacial binding. Quinone can polymerize with up to 6 DOPA residues (Fig. 6.2c) [62]. Additionally, quinone forms covalent cross-links with various functional groups (i.e., $-NH_2$, $-SH$, imidazole) found on biological substrates resulting in the formation of interfacial covalent cross-links (Fig. 6.2d) [40,63,64]. Catechols that contain a free amine group (i.e., dopamine) undergo intra-molecular cyclization between the catechol side chain and the amine group to form a leukochrome that can undergo polymerization similar to melanin formation [62]. Recently, the use of polydopamine as an adhesive primer for further grafting of antifouling polymers through a simple dip-coating process in an aqueous medium was reported [65].

6.3 Architectures of Synthetic Adhesive Polymers

Modification of DOPA and its catecholic derivatives onto inert, synthetic polymers readily impart these materials with the ability to cure through oxidant-induced cross-linking and to form strong, water-resistant interfacial bonds. This section reviews the two general architectures that have been widely used in designing biomimetic adhesive polymers: (1) main chain functionalization, and (2) end group functionalization (Fig. 6.3). It is possible to tune the materials and physical properties of the adhesive polymers for a desired application through incorporating the adhesive moiety in the polymers with various compositions and architectures.

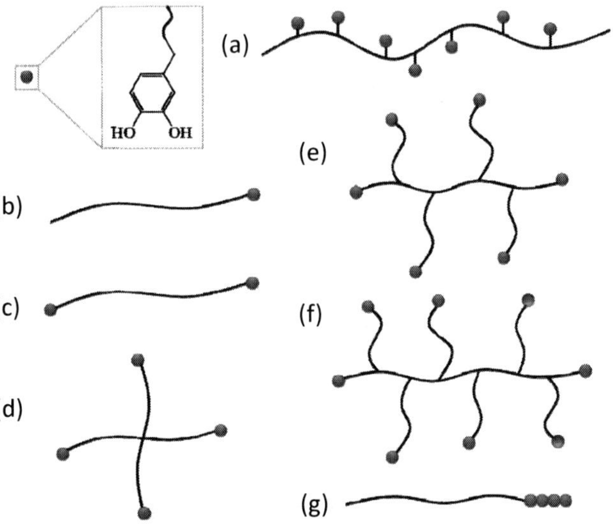

Figure 6.3 Chemical structures of synthetic polymers modified with catechol. Linear polymer with pendent catechol groups on main chain (a). Linear mono- (b) and bi-functional (c) polymers and branch 4- (d), 6- (e), and 8-armed (f) polymers end-capped with catechol. Linear polymer end-modified with oligomeric poly(DOPA) peptide (g).

6.3.1 Main Chain Functionalization

Natural adhesive proteins extracted from mussels are linear polymers containing pendant catechol groups functionalized on the main chain of the adhesive (Fig. 6.3a). Synthetic polymers can achieve a similar linear architecture through polymerization of catechol-containing monomers or through covalently linking catechol groups to the main chain of a linear polymer. Water soluble random co-polypeptide consisting of DOPA and lysine have been prepared by *N*-carboxyanhydride polymerization of these amino acids to yield adhesives with high molecular weights (MW >10^5 Da) and bonding strength [3,66]. Similarly, dopamine methacrylamide (DMA) consists of an adhesive moiety directly coupled to a polymerizable methacrylate group. DMA can be copolymerized with various monomer types (i.e. hydrophilic, hydrophobic, charged, zwitterionic) to prepare polymers with various physical properties, molecular weights (10^3–10^6 Da), and DMA contents (3–33 wt%) [67,68]. Styrene-based polymers have also been reported through copolymerization of 3,4-dihydroxystyrene and styrene or styrene sulfonate [69–71]. Finally, catechols in the form of DOPA or dopamine have been covalently attached to biopolymers such as hyaluronic acid [72], chitosan [73], alginate [74], heparin [75] and gelatin [67,68], as well as synthetic homopolymer or block copolymers composed of polyethylene glycol (PEG) and poly(caprolactone) (PCL) [67,68,76], as well as poly(methacrylate) [77].

6.3.2 End Group Functionalization

Adhesive catechol have been functionalized as end groups in linear mono- or bi-functional PEGs [16,78] and branched PEGs with 4-, 6-, and 8-arms [62,79] (Fig. 6.3b–f). Additionally, catechols have been modified onto linear or 4-armed polypropylene oxide (PPO)-PEG block copolymers that have demonstrated temperature responsiveness [80,81]. The reactive catechol group end-caps the water soluble and free moving PEG chain, which make it readily accessible for intermolecular cross-linking. Accessibility of the catechol for cross-linking is essential for rapid curing and formation of interfacial bonds for adhesion. Branched PEG can be linked with linear bifunctional PCL to form block copolymers with elevated MW (50–160 kDa) and materials properties due to hydrophobic–hydrophobic interactions between PCL segments [12,82,83]. In addition to end-modification with a single catechol group, linear PEG modified with oligomeric poly(DOPA) (Fig. 6.3g) [17,84], poly(DOPA-Lysine) [24,26], or MAP-mimic peptide sequences [16] have demonstrated enhanced adhesive properties to substrate surfaces. Similarly, branched PEGs end capped with oligomeric poly(DOPA) and poly(DOPA-lysine) peptides have resulted in drastically enhanced curing rate due to elevated DOPA content [85].

6.4 Bioinspired Synthetic Adhesives

Tissue adhesives can potentially simplify complex procedures, reduce surgery time, and minimize trauma typically associated with the use of sutures. However, commercially available adhesives are hampered by weak adhesive strength (e.g., fibrin glue) [86–88] and poor biocompatibility (e.g., cyanoacrylate-based adhesive) [89,90]. MAPs are excellent water-resistant adhesives and catechol's ability to undergo rapid curing and to form interfacial bonds with tissues has led to the development of various injectable tissue adhesives. Additionally, adhesives can be formulated into a thin adhesive film coated on a synthetic or biologic graft for soft tissue reinforcement and repair. Finally, hybrid adhesives that combine gecko-mimetic nanoscaled pillars with a mussel-mimetic adhesive coating have demonstrates reversible adhesion in both wet and dry conditions.

6.4.1 Injectable Tissue Adhesive

One of the earliest forms of synthetic mussel-mimetic adhesives are composed of catechol-modified branched PEG [13,62], due to the well-established biocompatibility of the polymer [91,92]. By controlling the PEG architecture, MW, catechol content, and linkage chemistry, PEG-catechol can be readily made with a tunable curing rate (instant to over an hour), degradation rate (days to over 3 months), degradation mechanism (hydrolysis vs. enzyme-mediated), extent of swelling (–20% to 300%), and controlled mechanical and adhesive properties [76,79,80,85]. These adhesives can be delivered using a dual barrel syringe applicator, which mixes the adhesive precursor with a mild oxidant (e.g., $NaIO_4$) in situ (Fig. 6.4). PEG-catechol adhesives have outperformed fibrin-based adhesives in various macroscopic adhesion tests (e.g., lap shear, burst strength, and peel, etc.) [76,79,85]. Additionally, these adhesives were demonstrated to be non-cytotoxic [93] and elicited minimal acute or chronic inflammatory response in vivo [76,94].

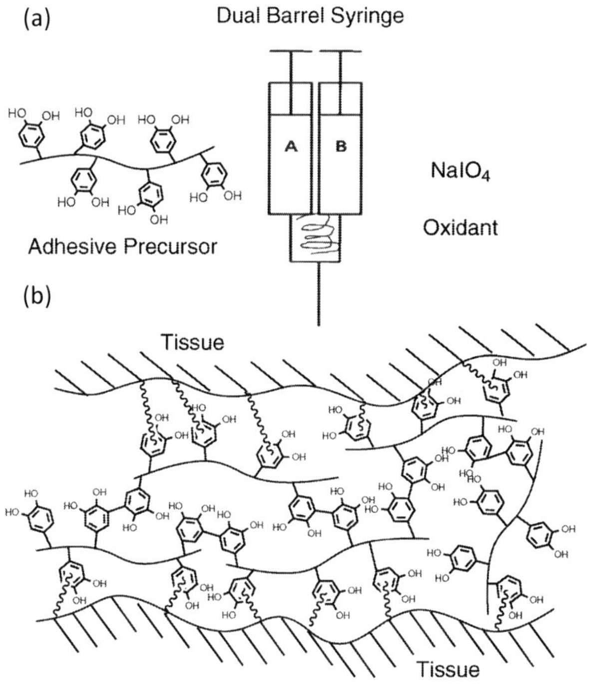

Figure 6.4 Schematic representation of injectable tissue adhesive delivered via a dual barrel syringe applicator. Adhesive precursor (a) is mixed with a mild oxidant such as sodium periodate (b) to form an adhesive hydrogel in situ.

The ability to tune the physical and materials properties of PEG-catechol adhesives while maintaining excellent biocompatibility is advantageous for tailoring this adhesive system to a variety of biomedical applications. An in vitro, side-by-side comparison of branched PEG-catechol with several other types of adhesive (cyanoacrylate-, fibrin-based adhesive, etc.) in sealing human fetal membrane defects revealed that only Tissucol (a fibrin glue) and the PEG-catechol adhesive showed efficient, non-disruptive, and nontoxic bonding [93]. However, fibrin glue needed dry membrane surfaces for reliable sealing, whereas PEG-catechol sealed a 3.7 mm fetal membrane defect up to 60 mbar (45 mmHg) when applied under wet conditions [95].

Recently, Mehdizadeh et al. [76] explored the use of PEG-catechol adhesives as a suture-less wound sealant. PEG-catechol achieved hemostasis instantly without the need of sutures to close a 2 cm-long by 0.5 cm-deep wound created on the back of Sprague–Dawley rats. Fibrin glue, the current gold standard, was not able to close this wound due to weak adhesion strength. Additionally, PEG-catechol adhesive completely degraded within 28 days and the mechanical properties of the adhesive-closed wound revealed better tensile strength when compared to that of suture-closed wounds.

In addition to functioning as a tissue adhesive or sealant, 4-armed PEG-catechol adhesive was utilized to immobilize islet cells, which remained functional in vivo for over a year [94]. The adhesive formed a translucent, thin hydrogel layer over the affixed islet bolus, resulting in islet entrapment between the adhesive and tissue surface. PEG-catechol immobilized islets were well vascularized after several months in vivo and actively secreted insulin. This result demonstrates the potential for using these biomimetic adhesives for cell and tissue engineering.

6.4.2 Nanocomposite Adhesive Hydrogel

While simple PEG-catechol adhesive hydrogels have demonstrated effective bioadhesive and physical properties that can be tailored to meet specific demands, they are fragile in nature due to elevated water content. One approach in enhancing the fracture resistance of hydrogels is to incorporate inorganic nanoparticles to form nanocomposite hydrogels, which have demonstrated elevated mechanical properties and extensibility as well as the ability to recover after large, repeated deformation [96,97]. These materials rely on weak physical bonds (e.g., hydrogen bonding, electrostatic interaction) formed between the polymer network and the encapsulated nanoparticles to increase the effective cross-linking density of the nanocomposite [98–100]. To enhance the interfacial binding between the polymer network and nanoparticles, DMA was copolymerized into a poly(acrylamide) nanocomposite hydrogel containing nanosilicate, Laponite, ($Na^{0.7+}(Mg_{5.5}Li_{0.3}Si_8)O_{20}(OH)_4)^{0.7-}$) [101]. DMA-containing nanocomposite demonstrated materials properties (e.g., maximum compressive strength, toughness, storage and loss moduli) that were an order of magnitude higher than DMA-free gels. DMA-containing nanocomposite hydrogels also demonstrated improved fracture resistance to compressive loading, capable of being repeatedly compressed to over 1 MPa without fracturing. The catechol side chain of DMA formed strong physical bonds with Laponite, which can dissipate fracture energy while minimizing permanent damage to the network architecture (Fig. 6.5).

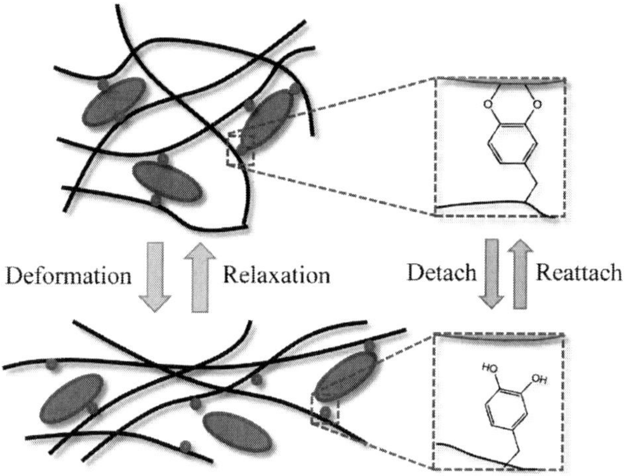

Figure 6.5 Schematic representation of a DMA-containing nanocomposite hydrogel. Deformation of the hydrogel breaks catechol-nanoparticle bonds, which reforms when the hydrogel relaxes [101]. Reproduced by permission of The Royal Society of Chemistry.

Laponite was formulated into a hydrogel made with 4-armed PEG end-capped with dopamine [102]. Dopamine's ability to form both irreversible covalent (cohesive and interfacial) and reversible physical (with Laponite) cross-links was exploited in creating an injectable tissue adhesive. Incorporation of Laponite increased the compressive materials properties (e.g., max strength, energy to failure, etc.) of the nanocomposite without compromising its compliance as strain at failure was also increased. From lap shear adhesion tests using wetted pericardium as the substrate, incorporating Laponite increased work of adhesion by fivefold over that of Laponite-free adhesive. Strong, physical bonds formed between dopamine and Laponite increased bulk materials properties, which contributed to the enhanced adhesive properties.

6.4.3 Thin-Film Adhesive

Another advance in improving the performance of biomimetic adhesives is the modification of amphiphilic multiblock copolymer films with the adhesive catechol [12,84,103,104]. These adhesive polymers contain PEG segments, which allow the polymer to remain relatively hydrophilic in order to achieve good "wetting" or adhesive contact with the surface substrate. Additionally, hydrophobic polyester segments, such as PCL, increase cohesive strength without creating a brittle material as they form reversible physical cross-links capable of dissipating localized fracture energy. These polymers can be cast as a thin adhesive film (50–120 μm) onto either biologic scaffolds (e.g., small intestinal submucosa, dermal tissues, and pericardium, Fig. 6.6) [12] or synthetic meshes [103,104] that are typically used in soft tissue repair (e.g., Achilles tendon, rotator cuff, hernia repair, etc.). These adhesive-coated constructs demonstrated adhesive strength that was 20 times higher than fibrin glue while reaching 60% that of cyanoacrylate-based adhesive.

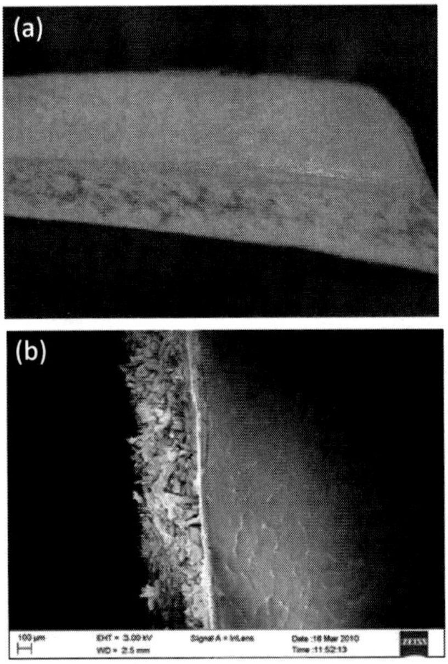

Figure 6.6 Photograph (a) and scanning electron microscopy image (b) of adhesive-coated biologic scaffold [11]. © IOP Publishing. Reproduced with permission.

Adhesive-coated scaffolds were used to augment primary suture repair of transected bovine Achilles tendons (Fig. 6.7) [11]. Ex vivo biomechanical testing of bioadhesive construct-augmented repair demonstrated significantly higher stiffness, failure load, and energy to failure when compared to those of sutured tendons (40%, 45%, and 63% increase over control, respectively). Higher load and energy was necessary to lengthen the repair and is potentially beneficial in preventing excessive elongation at the repair site, which leads to poor functional outcome after tendon repair [105,106]. Additionally, a stiffer repair also strongly correlated with increased failure properties of repaired tendons during the early phase of healing [106]. This study demonstrated that the adhesive construct augmentation can be utilized to reinforce primary suture repair and may lead to novel bioadhesive medical devices.

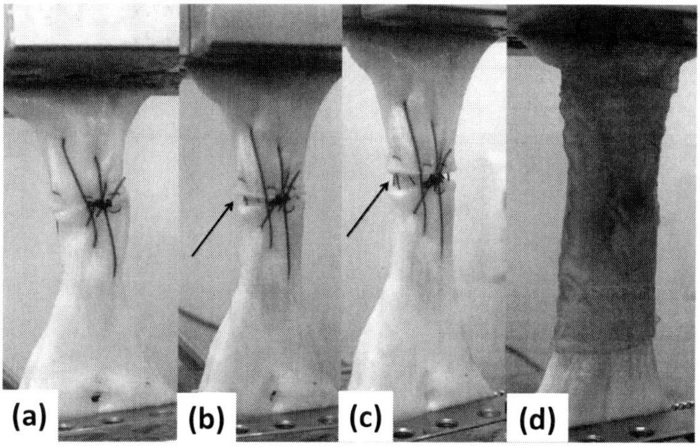

Figure 6.7 Photographs of sutured Achilles tendons loaded to 0 N (a), 50 N (b), and 100 N (c), and sutured tendon augmented with bioadhesive construct loaded to 100 N (d). Arrows indicate gap formation for tendons repaired with sutures [11]. © IOP Publishing. Reproduced with permission.

6.4.4 Hybrid Gecko- and Mussel-Inspired Adhesive

Scientists have attempted to recreate nanoscaled features that mimic the reversible adhesive properties of hierarchical gecko foot pads that allow these organisms to cling onto vertical and even inverted surfaces [107,108]. However, gecko-mimetic adhesion is greatly diminished upon full immersion in water [109,110]. To enhance wet adhesion, gecko-mimetic nanopillars were coated with a thin layer of mussel-mimetic random copolymer of DMA and 2-methoxyethyl acrylate (p(DMA-co-MEA)) (Fig. 6.8) [23]. p(DMA-co-MEA)-coated nanopillars demonstrated nearly 15-fold increase in adhesive properties in the presence of moisture while maintaining its adhesive performance for over 1000 contact cycles in both dry and wet environments. This hybrid adhesive can be further enhanced though optimizing the geometry of the nanostructure [111] as well as cross-linking of p(DMA-co-MEA) coating [112]. This unique combination may prove useful for reversible attachment to a variety of surfaces in many environments.

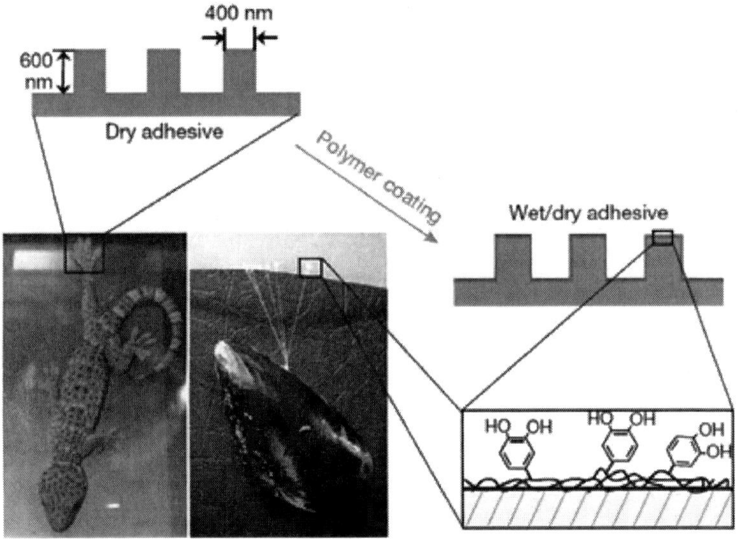

Figure 6.8 Gecko- and mussel-inspired hybrid nanostructure material shows robust adhesive properties both under dry and wet conditions. Reproduced from reference [23] with permission.

6.5 Mussel-Inspired Antifouling Coatings

Control of interfacial adsorption events plays a critical role in the success of many applications ranging from implantable medical devices to seafaring vessels [113]. Left uncontrolled, non-specific adsorption of biomolecules, bacteria, and other organisms on device surfaces severely impairs function. One strategy in minimizing fouling on a surface is to modify it with antifouling polymers such as PEG, which has successfully reduced the interfacial adsorption of cells [16,26,114,115], microbes [84,116–118] and macromolecules [18,119]. The hydrophilic nature of these polymers strongly couples water molecules when tethered to a surface, thus limiting the access of approaching cells and macromolecules to adsorption sites due to associated thermodynamic (entropic and enthalpic) penalties (i.e., steric hindrance). Many anchoring strategies involving the formation of robust chemical bonds between the polymer and surface exist, as exemplified by thiol–Au [120] or silane–metal oxide interactions [121]. However, these approaches suffer from significant limitations such as substrate chemical specificity [120,121], complicated and expensive protocols [122,123], and susceptibility to thermal or hydrolytic degradation [124,125].

The ability to adhere to both organic and inorganic surfaces using a simple dip-coating approach makes catechol a versatile anchoring group for tethering antifouling polymers. Surface modification has been demonstrated using two grafting approaches: (1) adsorption of a polymer onto a substrate from solution (so-called "graft-to" approach) and (2) surface-initiated polymerization (SIP) of monomers from a substrate-bound initiator ("graft-from") (Fig. 6.9).

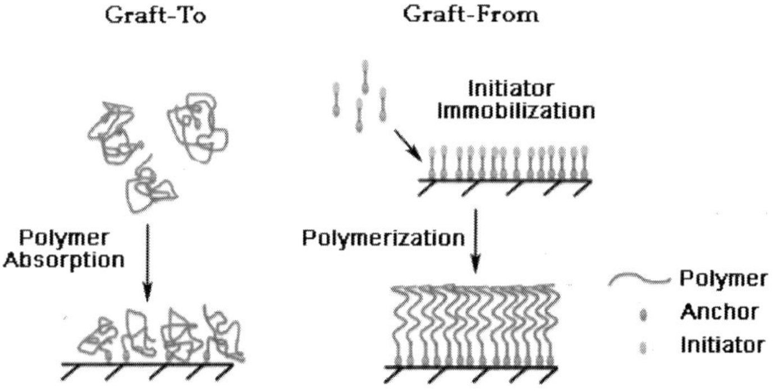

Figure 6.9 Graft-to vs. graft-from coating methods used in preparing antifouling surfaces.

6.5.1 Graft-to Coating Method

The graft-to coating approach involves the adsorption of preformed linear antifouling polymer end-functionalized with a catechol anchoring group, such as methoxy-PEG-DOPA on to a surface

[16]. Further enhancement in interfacial binding strength can be achieved using a short peptide consisting of poly(DOPA) [17], poly(DOPA-Lysine) [26], and peptides that mimic MAP sequences [16] as the anchor or using main chain-functionalized polymers with high MW [67,68,82,83]. These polymers have rendered metallic (Au_2O_3, Al_2O_3, SiO_2, TiO_2, NiTi, and stainless steel) and polymeric (polyurethane and silicone rubber) substrates resistant to fouling from proteins, cells, bacteria, and algae [16–18,25,26, 84,126].

Due to steric hindrance of adsorbing polymers with elevated MW, the graft-to approach can suffer from low grafting density with reduced antifouling performance [127]. To develop coatings with sufficient polymer densities, researchers have utilized the cloud point grafting approach, which exploits the lower critical solution temperature (LCST) of the polymer of interest [18]. At a temperature just below LCST or the cloud point, polymer chains are only marginally soluble and possess a greatly reduced hydrodynamic radius, which permits higher packing densities on surfaces.

Earlier development of antifouling technology has focused on PEG-based coatings, due to its well-established effectiveness and biocompatibility. However, PEG can be degraded through oxidation [123,128,129], which can compromise its long-term effectiveness. Messersmith and colleagues [130–132] synthesized a series of linear poly(*N*-substituted glycine) or α-peptoid to form an antifouling coating that repels non-specific binding of proteins, cells, and bacteria. A peptoid has a peptide-like polyamide backbone with side chain substitution on the amide nitrogen instead of the α-carbon, which make peptoids more protease-resistant than peptides [133]. Other polymers such as ionic and zwitterionic polymers have demonstrated promising results [67,68,134,135]. Finally, catechol-modified biopolymers have been used to enhance hemocompatibility [75] and promote specific binding of stem cells [72].

6.5.2 Graft-From Coating Method

The graft-from approach utilizes catechol groups to anchor the initiator onto the surface to perform surface-initiated polymerization to produce a dense layer of polymer brushes [127]. Dopamine functionalized with alkyl bromine has initiated poly(methacrylate) brushes with oligomeric ethylene glycol side chains using surface-initiated atom transfer radical polymerization (SI-ATRP) [136,137]. SI-ATRP chemistry is facile and highly tunable. When combined with photolithography techniques, surface patterns of various shapes (e.g., squares, lines) in the micron-scale were created [137,138]. Additionally, SI-ATRP can be used to graft poly(methacrylate) polymers with various side chain functional groups, including oligomeric ethylene glycol (OEG) [136,137], zwitterionic carboxybetaine [139], and thermo-responsive *N*-isopropylacrylamide [138]. These coatings have rendered surfaces such as titanium and gold highly resistant to fouling from cells and proteins [136,139]. Surface-initiated polymerization not only can be used to tether polymers onto planar substrates, but to nanoparticles as well [137,138], which further expands its application in designing novel materials. In addition to ATRP chemistry, Ye et al. [140] utilized surface-initiated ring-opening metathesis polymerization SI-ROMP chemistry to graft poly (pentadecafluorooctyl-5-norbornene-2-carboxylate) brushes onto TiO_2 nanowires to create a superhydrophobic surface capable of repelling water and alkane liquids.

6.6 Polydopamine as a Versatile Multifunctional Anchoring Group

Dopamine (Fig. 6.1b) consists of a catechol and an amine group, which mimics the side chain functional groups of DOPA and lysine, respectively. Both amino acids are found in large abundance in *Mytilus edulis* foot protein-5 (*Mefp-5*), which is found predominately at the adhesive plaque-substrate interface and is believed to have an interfacial role [141]. Oxidation of dopamine leads to intra-molecular cyclization to form a leukochrome which can further polymerize through reaction pathways similar to melanin formation [65]. The polymerized dopamine (polydopamine) remains reactive and can be further modified with polymers, initiators, proteins, polysaccharides, oligonucleotides, nanoparticles, growth factors, cells, or metal ions (Fig. 6.10) [65,142–146].

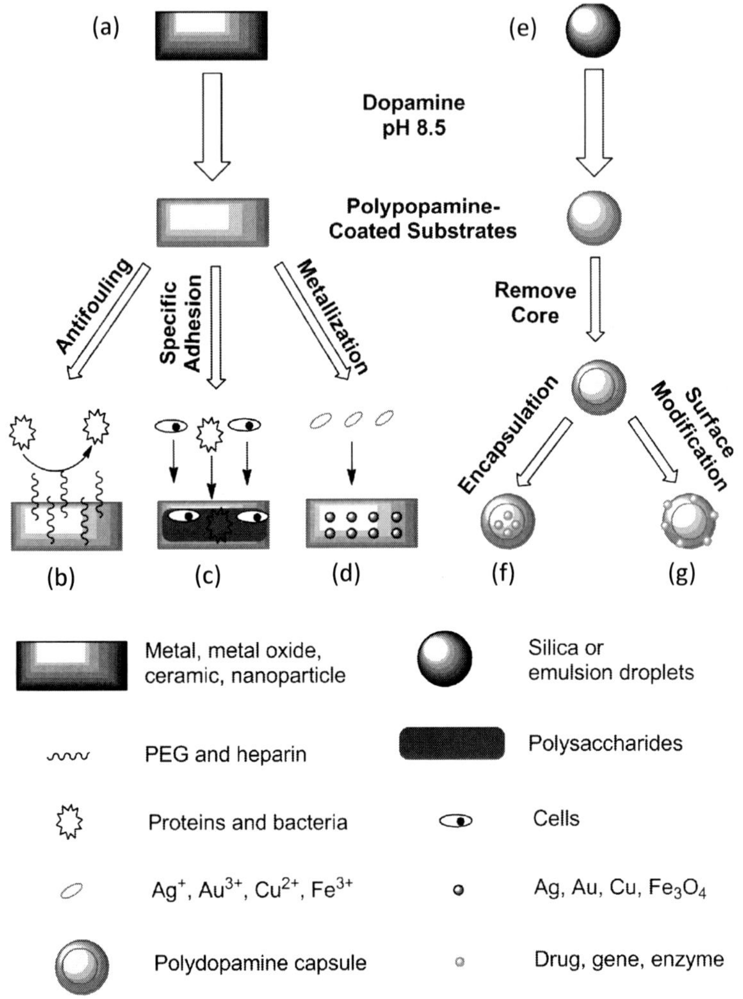

Figure 6.10 Schematic representation of polydopamine coating and subsequent functionalization. Substrates are immersed in a dopamine solution (pH 8.5) for several hours to form a polydopamine coated surface (a), which can be further modified with antifouling polymers (b), polysaccharides for specific cell adhesion (c), and metal ions to form metallized coating (d). When polydopamine is coated onto a template (e), removal of the template results in the formation of polydopamine capsule for carrying active compounds either encapsulated within the lumen (f) or tethered to the surface (g) of the capsule.

6.6.1 Polydopamine-Mediated Surface Modification

The polydopamine coating process involves a facile two-step process: (1) adsorption and polymerization of dopamine, and (2) functionalization of the polydopamine coating. When dopamine polymerizes in the presence of a material or a surface, polydopamine is coated on these substrates forming a thin layer (5–50 nm) of primer. The coating process utilizes a simple dip-coating approach in an aqueous (pH 8.5) solution that can be performed on a wide range of surfaces, ranging from metals, metal oxides, semiconductors, ceramics, and polymers (Fig. 6.10a) [65]. When combined with photolithography techniques, polydopamine can be coated into micron-scaled patterns. Coating onto nanoparticles, nanowires, and nanosheets have also been demonstrated [145,147,148]. Due to the mild coating process, researchers have demonstrated using polydopamine coatings to individually encapsulate yeast cells, which maintained their viability within polydopamine shells and have controlled the cell cycle by the thickness of the coating [143]. Similarly, polydopamine could encapsulate *Escherichia coli* to inhibit its growth to achieve antibacterial activity [149].

Polydopamine coatings readily react with polymers, proteins, or biomolecules bearing functional groups such as amine, thiol, or imidazole groups, covalently linking these polymers to the polydopamine-coated surface [65,142]. Polydopamine coating demonstrated pH dependent selectivity in binding to imidazole ($pK_a \sim 6$) and primary amine ($pK_a \sim 10$), depending on the protonation of the reactive group [142]. Grafting with antifouling polymers such as PEG (Fig. 6.10b) have rendered the surface antifouling while modification with polysaccharides (Fig. 6.10c), such as hyaluronic acid, have promoted specific binding of bone marrow and stem cells [65]. Although polydopamine-coated surfaces have demonstrated improved blood compatibility and suppressed palate adhesion [150], further treatment with heparin has markedly improved hemocompatibility of graphene nanosheets [147]. Recently, Kang et al. [144] demonstrated that one-pot surface modification is feasible when the molecules (e.g., polymers, polysaccharides, initiators, etc.) to be tethered are co-dissolved with dopamine as it undergoes polymerization, which further simplifies the functionalization process.

6.6.2 Metallization and Metal Ion Reduction

Catechol's affinity with metal ions has been exploited to metallize polydopamine coatings through incubating the polydopamine-coated objects in solutions containing these ions (Ag^+, Au^+, Cu^{2+}, Fe^{3+}, etc.) (Fig. 6.10d) [65,148,151,152]. During the process, nanoparticles nucleate on the polydopamine coating. This process of preparing nanocomposite material does not require harsh reaction conditions such as high temperature and pressure, hazardous chemicals, or reducing agents. More importantly, the nanoparticles could be adsorbed onto the surface without utilization of capping agents [151].

One sample application of a nanoparticle-modified polydopamine coating is the use of generating an antibacterial surface with attached silver nanoparticles [152]. Silver is well known for its non-specific antimicrobial properties for the treatment of wounds, burns, and bacterial infections [153]. Compared to bulk silver and silver solutions, silver nanoparticles are tunable for commercial medicinal material modification such as nanocomposite antibacterial gels [154] and nanoparticle incorporating coatings [155]. Silver nanoparticle modified catheters have shown antimicrobial properties both in vitro and in vivo studies [156]. In order to reduce the toxic release of silver nanoparticles, Sileika et al. [157] further grafted PEG to enhance the antifouling property of the surface.

6.6.3 Polydopamine Capsule

Polydopamine has been utilized to form an organic shell coated onto living cells [143], magnetic Fe_3O_4 particles, fluorescent quantum dots, or anti-cancer drugs [158]. When polydopamine is formed on a temporary template (e.g., emulsion droplets or acid degradable SiO_2), the removal of the template core results in the formation of a polydopamine capsule (Fig. 6.10e) [158,159]. Preloading the template with a functional substance (Fig. 6.10f) such as drug particles can be used to deliver these molecules in a controlled manner by tailoring the permeability of the capsule based on its thickness and size [158]. Additionally, polydopamine capsule surfaces can be further functionalized with pH- and temperature-sensitive polymer brushes to further modulate the release of drug particles with changes to the environment [159]. Polydopamine capsules have also been used to mimic the catalytic function of mitochondria [160]. Three enzymes (α-amylase, β-amylase and glucosidase) were separately immobilized in the lumen, within the wall, and on the outer surface of the capsule (Fig. 6.10g). The unique structural organization of this multi-enzyme system displayed higher catalytic activity and enhanced operational stability.

6.7 Summary and Future Outlook

MAPs–inspired technologies are motivated by the exceptional adhesive properties of these proteins and the need to overcome moisture in developing synthetic adhesives and coatings. Unlike other biomimetic approaches where a peptide sequence is necessary to mimic the function of a naturally derived protein (e.g., arginine-glycine-aspartic acid for cellular adhesion), incorporation of a single DOPA amino acid or its derivatives can impart inert polymer systems with strong, water-resistant adhesive properties. The simplicity and the wide range of chemistry that catechols can participate in have spawned a large volume of literature aiming at harvesting these adhesive properties for a wide range of applications. While this chapter predominately reviewed works that focused on applications in the biomedical field, this technology can potentially be applied to other industries as well. Wilker and colleagues [71] have developed adhesives with a styrene-based backbone, which may be seamlessly integrated into the production of plastic materials, whether it is for interfacial bonding or improving materials properties. Similarly, Pan et al. [161] is exploring the use of styrene-butadiene end-capped with catechol for developing nanocomposite rubbers for improved performance in tires. Additionally, polydopamine coatings have been exploited to remove heavy metal ions from drinking water [162] and controlled release of fertilizer [163].

Existing synthetic mimics of MAPs have exclusively focused on exploiting the chemistry of the DOPA side chain. However, mussels rely on numerous ingenious designs found in both the adhesive plaque and byssus thread to resist repeated pounding by intertidal waves. Histidine-rich domains are found in byssus threads, which form reversible bonds with transitional metal ions (zinc and copper) [34]. Histidine–metal ion bonds can break and reform while minimizing permanent structural damage. This interaction was recently exploited to create a self-healing hydrogel [164]. The byssus thread also exhibits a stiffness gradient with a modulus of 900 MPa at the distal end and 20 MPa near the mussel tissue [165]. Similarly, the adhesive plaque exhibits a gradient in pore diameters, ranging from 200 nm near the substratum but nearly 3 μm where the thread meets the plaque [166]. Gradation in the materials properties may be nature's adaptation to prevent structural damage due to contact deformation between two dissimilar materials [167]. These structural designs provide new insights in developing novel adhesives with enhanced interfacial properties.

Acknowledgment

The authors would like to acknowledge National Institutes of Health (GM104846) and Michigan Technological University (1205015P1) for financial support. MMC was partially supported by Environmental Protection Agency Greater Research Opportunities Undergraduate Fellowship.

References

1. Comyn, J. (1981). The relationship between joint durability and water diffusion, in *Developments in Adhesives* (eds., Kinloch, A. J.), Applied Science Publishers, Barking, UK, pp. 279–313.
2. Waite, J. H. (1987). Nature's underwater adhesive specialist, *Int. J. Adhes. Adhes.*, **7**, 9–14.
3. Yu, M., Hwang, J., and Deming, T. J. (1999). Role of L-3,4-dihydroxyphenylanine in mussel adhesive proteins, *JACS*, **121**, 5825–5826.
4. Deming, T. J., Yu, M., and Hwang, J. (1999). Mechanical studies of adhesion and crosslinking in marine adhesive protein analogs, *Polym. Mater. Sci. Eng.*, **80**, 471–472.
5. Waite, J. H. (1991). Mussel beards: A coming of age, *Chem. Ind.*, **2**, 607–611.
6. Lee, B. P., Messersmith, P. B., Israelachvili, J. N., and Waite, J. H. (2011). Mussel-inspired adhesives and coatings, *Ann. Rev. Mater. Res.*, **41**, 99–132.

7. Lee, B. P., Dalsin, J. L., and Messersmith, P. B. (2006). Biomimetic adhesive polymers based on mussel adhesive proteins, in *Biological Adhesives* (eds., Smith, A. M., and Callow, J. A.), Springer Berlin Heidelberg, Berlin, Germany, pp. 257–278.
8. Waite, J. H., and Andersen, S. O. (1978). 3,4-Dihydroxyphenylalanine in an insoluble shell protein of *Mytilus edulis*, *Biochim. Biophys. Acta.*, **541**, 107–114.
9. Ninan, L., Monahan, J., Stroshine, R. L., Wilker, J. J., and Shi, R. (2003). Adhesive strength of marine mussel extracts on porcine skin, *Biomaterials*, **24**, 4091–4099.
10. Burzio, L. O., Burzio, V. A., Silva, T., Burzio, L. A., and Pardo, J. (1997). Environmental bioadhesion: Themes and applications, *Curr. Opin. Biotechnol.*, **8**, 309–312.
11. Brodie, M., Vollenweider, L., Murphy, J. L., Xu, F., Lyman, A., Lew, W. D., and Lee, B. P. (2011). Biomechanical properties of Achilles tendon repair augmented with bioadhesive-coated scaffold, *Biomed. Mater.*, **6**, 015014.
12. Murphy, J. L., Vollenweider, L., Xu, F., and Lee, B. P. (2010). Adhesive performance of biomimetic adhesive-coated biologic scaffolds, *Biomacromolecules*, **11**, 2976–2984.
13. Burke, S. A., Ritter-Jones, M., Lee, B. P., and Messersmith, P. B. (2007). Thermal gelation and tissue adhesion of biomimetic hydrogels, *Biomed. Mater.*, **2**, 203–10.
14. Huang, K., Lee, B. P., Ingram, D., and Messersmith, P. B. (2002). Synthesis and characterization of self-assembling block copolymers containing bioadhesive end groups, *Biomacromolecules*, **3**, 397–406.
15. Shao, H., Bachus, K. N., and Stewart, R. J. (2009). A water-borne adhesive modeled after the sandcastle glue of *P. californica*, *Macromol. Biosci.*, **9**, 464–471.
16. Dalsin, J. L., Hu, B.-H., Lee, B. P., and Messersmith, P. B. (2003). Mussel adhesive protein mimetic polymers for the preparation of nonfouling surfaces, *JACS*, **125**, 4253–4258.
17. Dalsin, J. L., Lin, L., and Messersmith, P. B. (2004). Antifouling performance of poly(ethylene glycol) anchored onto surfaces by mussel adhesive protein mimetic peptides, *Polym. Mater. Sci. Eng.*, **90**, 247–248.
18. Dalsin, J. L., Lin, L., Tosatti, S., Voeroes, J., Textor, M., and Messersmith, P. B. (2005). Protein resistance of titanium oxide surfaces modified by biologically inspired mPEG-DOPA, *Langmuir*, **21**, 640–646.

19. Araujo, P. Z., Morando, P. J., and Blesa, M. A. (2005). Interaction of catechol and gallic acid with titanium dioxide in aqueous suspensions. 1. equilibrium studies, *Langmuir*, **21**, 3470–3474.

20. Creutz, C., and Chou, M. H. (2008). Binding of catechols to mononuclear titanium (IV) and to 1- and 5-nm TiO_2 nanoparticles, *Inorg. Chem.*, **47**, 3509–3514.

21. Kawabata, T., Schepkin, V., Haramaki, N., Phadke, R. S., and Packer, L. (1996). Iron coordination by catechol derivative antioxidants, *Biochm. Pharmacol.*, **51**, 1569–1577.

22. Amstad, E., Gillich, T., Bilecka, I., Textor, M., and Reimhult, E. (2009). Ultrastable iron oxide nanoparticle colloidal suspensions using dispersants with catechol-derived anchor groups, *Nano Lett.*, **9**, 4042–4048.

23. Lee, H., Lee, B. P., and Messersmith, P. B. (2007). A reversible wet/dry adhesive inspired by mussels and geckos, *Nature*, **448**, 338–341.

24. Chawla, K., Lee, S., Lee, B. P., Dalsin, J. L., Messersmith, P. B., and Spencer, N. D. (2009). A novel low-friction surface for biomedical applications: Modification of poly(dimethylsiloxane) (PDMS) with polyethylene glycol(PEG)-DOPA-lysine, *J. Biomed. Mater. Res. Part A.*, **90A**, 742–749.

25. Pechey, A., Elwood, C. N., Wignall, G. R., Dalsin, J. L., Lee, B. P., Vanjecek, M., Welch, I., Ko, R., Razvi, H., and Cadieux, P. A. (2009). Anti-adhesive coating and clearance of device associated uropathogenic escherichia coli cystitis, *J. Urol.*, **182**, 1628–1636.

26. Dalsin, J. L., Sherman, D. L., Lee, B. P., and Messersmith, P. B. (2006). Versatile mussel adhesive-inspired biomimetic antifouling polymers, *PMSE Preprints.*, **94**, 854–855.

27. Coombs, T. L., and Keller, P. J. (1981). Mytilus byssal threads as an environmental marker for metals, *Aquat. Toxicol.*, **1**, 291–300.

28. Szefer, P., Ikuta, K., Kushiyama, S., Szefer, K., Frelek, K., and Geldon, J. (1997). Distribution and association of trace metals in soft tissue and byssus of *Mytilus edulis* from the east coast of Kyushu Island, Japan, *Arch Environ. Contam. Toxicol.*, **32**, 184–190.

29. Martin, S., Kesselman, J., Park, D., Lewis, N., and Hoffmann, M. (1996). Surface studies of 4-chlorocatechol adsorbed on titanium dioxide, *Environ. Sci. Technol.*, **30**, 2535–2542.

30. Sever, M. J., and Wilker, J. J. (2004). Visible absorption spectra of metal-catecholate and metal-tironate complexes, *Dalton Trans.*, **7** 1061–1072.

31. Taylor, S. W., Chase, D. B., Emptage, M. H., Nelson, M. J., and Waite, J. H. (1996). Ferric ion complexes of a DOPA-containing adhesive protein from *Mytilus edulis*, *Inorg. Chem.*, **35**, 7572–7577.

32. Taylor, S. W., Luther, G. W., III, and Waite, J. H. (1994). Polarographic and spectrophotometric investigation of iron(III) complexation to 3,4-dihydroxyphenylalanine-containing peptides and proteins from *Mytilus edulis*, *Inorg. Chem.*, **33**, 5819–5824.
33. Harrington, M. J., Masic, A., Holten-Andersen, N., Waite, J. H., and Fratzl, P. (2010). Iron-clad fibers: A metal-based biological strategy for hard flexible coatings, *Science*, **328**, 216–220.
34. Harrington, M. J., Gupta, H. S., Fratzl, P., and Waite, J. H. (2009). Collagen insulated from tensile damage by domains that unfold reversibly: In situ X-ray investigation of mechanical yield and damage repair in the mussel byssus, *J. Struct. Biol.*, **167**, 47–54.
35. Holten-Andersen, N., Harrington, M. J., Birkedal, H., Lee, B. P., Messersmith, P. B., Lee, K. Y. C., and Waite, J. H. (2011). pH-Induced mussel metal-ligand crosslinks yield self-healing polymer networks with near-covalent elastic moduli, *Proc. Nat. Acad. Sci.*, **15**, 2651–2655.
36. Kummert, R., and Stumm, W. (1980). The surface complexation of organic acids on hydrous g-alumina, *J. Colloid Interface Sci.*, **75**, 373–385.
37. Soriaga, M. P., and Hubbard, A. T. (1982). Determination of the orientation of aromatic molecules adsorbed on platinum electrodes. The effect of solute concentration, *JACS*, **104**, 3937–3945.
38. Hansen, D. C., Dexter, S. C., and Waite, J. H. (1995). The inhibition of corrosion of S30403 stainless steel by a naturally occurring catecholic polymer, *Corros. Sci.*, **37**, 1423–1441.
39. Lee, B. P., Chao, C.-Y., Nunalee, F. N., Motan, E., Shull, K. R., and Messersmith, P. B. (2006). Rapid photocurable of amphiphilic block copolymers hydrogels with high DOPA contents, *Maclomolecules*, **39**, 1740–1748.
40. Lee, H., Scherer, N. F., and Messersmith, P. B. (2006). Single-molecule mechanics of mussel adhesion, *Proc. Natl. Acad. Sci. U. S. A.*, **103**, 12999–13003.
41. Mian, S. A., Gao, X., Nagase, S., and Jang, J. (2011). Adsorption of catechol on a wet silica surface: Density functional theory study, *Theor. Chem. Acc.*, **130**, 333–339.
42. Mian, S. A., Saha, L. C., Jang, J., Wang, L., Gao, X., and Nagase, S. (2010). Density functional theory study of catechol adhesion on silica surfaces, *J. Phys. Chem. C.*, **114**, 20793–20800.
43. Vega-Arroyo, M., LeBreton, P. R., Rajh, T., Zapol, P., and Curtiss, L. A. (2005). Density functional study of the TiO_2–dopamine complex, *Chem. Phys. Lett.*, **406**, 306–311.

44. Deacon, M. P., Davis, S. S., Waite, J. H., and Harding, S. E. (1998). Structure and mucoadhesion of mussel glue protein in dilute solution, *Biochem.*, **37**, 14108–14112.

45. Schnurrer, J., and Lehr, C.-M. (1996). Mucoadhesive properties of the mussel adhesive protein, *Int. J. Pharm.*, **141**, 251–256.

46. Chirdon, W. M., O'Brien, W. J., and Robertson, R. E. (2003). Adsorption of catechol and comparative solutes on hydroxyapatite, *J. Biomed. Mat. Res. B*, **66B**, 532–538.

47. Baty, A. M., Leavitt, P. K., Siedlecki, C. A., Tyler, B. J., Suci, P. A., Marchant, R. E., and Geesey, G. G. (1997). Adsorption of adhesive proteins from the marine mussel, *Mytilus edulis*, on polymer films in the hydrated state using angle dependent X-ray photoelectron spectroscopy and atomic force microscopy, *Langmuir*, **13**, 5702–5710.

48. Baker, B. R., Laiwalla, A. N., Yoon, J.-Y., Canavate, J., and Garrell, R. L. (2001). Adhesion and cohesion of mussel adhesive protein on glass and gold through protein removal studies, *Polym. Mater. Sci. Eng.*, **85**, 115–116.

49. Litvinov, R. I., Bennett, J. S., Weisel, J. W., and Shuman, H. (2005). Multi-step fibrinogen binding to the integrin αIIbβ3 detected using force spectroscopy *Biophys. J.*, **89**, 2824–2834.

50. Kienberger, F., Kada, G., Gruber, H. J., Pastushenko, V. P., Riener, C., Trieb, M., Knaus, H.-G., Schindler, H., and Hinterdorfer, P. (2000). Recognition force spectroscopy studies of the NTA-His6 bond, *Single Mol.*, **1**, 59–65.

51. Florin, E. L., Moy, V. T., and Gaub, H. E. (1994). Adhesion forces between individual ligand-receptor pairs, *Science*, **264**, 415–417.

52. Wong, S. S., Joselevich, E., Woolley, A. T., Cheung, C. L., and Lieber, C. M. (1998). Covalently functionalized nanotubes as nanometre-sized probes in chemistry and biology, *Nature*, **394**, 52–54.

53. Oesterhelt, F., Rief, M., and Gaub, H. E. (1999). Single molecule force spectroscopy by AFM indicates helical structure of poly (ethylene-glycol) in water, *N. J. Phys.*, **1**, 1–11.

54. Andersen, S. O., Jacobsen, J. P., Bojesen, G., Paul Hedlin, and Roepstorff, P. (1992). Phenoloxidase catalyzed coupling of catechols. Identification of novel coupling products, *Biochim. Biophys. Acta.*, **1118**, 134–138.

55. Kramer, K. J., Morgan, T. D., Hopkins, T. L., Christensen, A., and Schaefer, J. (1991). Insect cuticle tanning: Enzymes and cross-link structure, in *Natural Occurring Pest Bioregulators*, eds. pp. 87–105.

56. Lee, B. P., Dalsin, J. L., and Messersmith, P. B. (2001). Enzymatic and non-enzymatic pathways to formation of DOPA-modifled PEG hydrogels, *Polym. Prepr.*, **42**, 151–152.

57. Yamamoto, H., and Tatehata, H. (1995). Oxidation of synthetic precursors of adhesive proteins from mytilid bivalves using tyrosinase, *J. Mar. Biotechnol.*, **2**, 95–100.

58. Rodgers, K. J., and Dean, R. T. (2000). Metabolism of protein-bound DOPA in mammals, *Int. J. Biochem. Cell Biol.*, **32**, 945–955.

59. Filpula, D. R., Lee, S. M., Link, R. P., Strausberg, S. L., and Strausberg, R. L. (1990). Structural and functional repetition in a marine mussel adhesive protein, *Biotechnol. Prog.*, **6**, 171–177.

60. Burzio, L. A., and Waite, J. H. (2000). Cross-linking in adhesive quinoproteins: Studies with model decapeptides, *Biochem.*, **39**, 11147–11153.

61. Herlinger, E., Jameson, R. F., and Linert, W. (1995). Spontaneous autoxidation of dopamine, *J. Chem. Soc. Perk. T.*, **2**, 259–263.

62. Lee, B. P., Dalsin, J. L., and Messersmith, P. B. (2002). Synthesis and gelation of DOPA-modified poly(ethylene glycol) hydrogels, *Biomacromolecules*, **3**, 1038–1047.

63. Waite, J. H. (1990). The phylogeny and chemical diversity of quinone-tanned glues and varnishes, *Comp. Biochem. Physiol. B*, **97**, 19–29.

64. Sugumaran, M., Dali, H., and Semensi, V. (1989). Chemical- and cuticular phenoloxidase-mediated synthesis of cysteinyl-catechol adducts, *Arch. Insect Biochem. Physiol.*, **11**, 127–137.

65. Lee, H., Dellatore, S. M., Miller, W. M., and Messersmith, P. B. (2007). Mussel-inspired surface chemistry for multifunctional coatings, *Science*, **318**, 426–430.

66. Yu, M., and Deming, T. J. (1998). Synthetic polypeptide mimics of marine adhesives, *Macromolecules*, **31**, 4739–4745.

67. Lee, B. P. (2009). Inventor biomimetic compounds and synthetic methods therefor. US patent 7,622,533.

68. Lee, B. P., inventor (2011). Biomimetic compounds and synthetic methods therefor. US Patent 8,030,413.

69. Daly, W. H., and Moulay, S. (1986). Synthesis of poly(vinylcatechols), *J. Polym. Sci.*, **74**, 227–242.

70. Yang, Z., and Pelton, R. (1998). The synthesis of poly(3,4-dihydroxystyrene) and poly[(sodium 4-styrenesulfonate)-co-(3,4-dihydroxystyrene)], *Macromol. Rapid Commun.*, **19**, 241–246.

71. Westwood, G., Horton, T. N., and Wilker, J. J. (2007). Simplified polymer mimics of cross-linking adhesive proteins, *Maclomolecules*, **40**, 3960–3964.

72. Hong, S., Yang, K., Kang, B., Lee, C., Song, I. T., Byun, E., Park, K. I., Cho, S.-W., and Lee, H. (2013). Hyaluronic acid catechol: A biopolymer exhibiting a pH-dependent adhesive or cohesive property for human neural stem cell engineering, *Adv. Funct. Mater.*, **23**, 1774–1780.

73. Lee, K., Oh, M. H., Lee, M. S., Nam, Y. S., Park, T. G., and Jeong, J. H. (2013). Stabilized calcium phosphate nano-aggregates using a dopa-chitosan conjugate for gene delivery, *Int. J. Pharm.*, **445**, 196–202.

74. Kastrup, C. J., Nahrendorf, M., Figueiredo, J. L., Lee, H., Kambhampati, S., Lee, T., Cho, S.-W., Gorbatov, R., Iwamoto, Y., Dang, T. T., Dutta, P., Yeon, J. H., Cheng, H., Pritchard, C. D., Vegas, A. J., Siegel, C. D., MacDougall, S., Okonkwo, M., Thai, A., Stone, J. R., Coury, A. J., Weissleder, R., Langer, R., and Anderson, D. G. (2012). Painting blood vessels and atherosclerotic plaques with an adhesive drug depot, *Proc. Nat. Acad. Sci.*, **109**, Issue, 52, pp. 21444–21449.

75. You, I., Kang, S. M., Byun, Y., and Lee, H. (2011). Enhancement of blood compatibility of poly(urethane) substrates by mussel-inspired adhesive heparin coating, *Bioconjug. Chem.*, **22**, 1264–1269.

76. Mehdizadeh, M., Weng, H., Gyawali, D., Tang, L., and Yang, J. (2012). Injectable citrate-based mussel-inspired tissue bioadhesives with high wet strength for sutureless wound closure, *Biomaterials*, **33**, 7972–7983.

77. Guvendiren, M., Brass, D. A., Messersmith, P. B., and Shull, K. R. (2009). Adhesion of DOPA-functionalized model membranes to hard and soft surfaces, *J. Adh.*, **86**, 631–645.

78. Zeng, X., Westhaus, E., Eberle, N., Lee, B. P., and Messersmith, P. B. (2000). Synthesis and characterization of DOPA-PEG conjugates, *Polym. Prepr.*, **41**, 989–990.

79. Dalsin, J. L., Lee, B. P. Vollenweider, L., Silvary, S., Murphy, J. L., Xu, F., Spitz, A., and Lyman, A., inventors (2012). Multi-armed catechol compound blends. US Patent 8,119,742.

80. Barrett, D. G., Bushnell, G. G., and Messersmith, P. B. (2013). Mechanically robust, negative-swelling, mussel-inspired tissue adhesives, *Adv. Healthcare Mater.*, **2**, 745–755.

81. Huang, K., Lee, B. P., Ingram, D., and Messersmith, P. B. (2001). Synthesis and characterization of self-assembling block copolymers containing adhesive moieties, *Polym. Prepr.*, **42**, 147–148.

82. Lee, B. P., Murphy, J. L., and Silvary, S., inventors (2010). Multibranched bioadhesive compounds and synthetic methods therefor patent, US Patent Application 20100197868.

83. Lee, B. P., Murphy, J. L., and Silvary, S. inventors (2011). Multi-linked star-shaped polymers and synthetic methods therefor patent, US Patent Application 20110318394.

84. Ko, R., Cadieux, P. A., Dalsin, J. L., Lee, B. P., Elwood, C. N., and Razvi, H. (2008). Novel uropathogen-resistant coatings inspired by marine mussels, *J. Endourol.*, **22**, 1153–1160.

85. Messersmith, P. B., Lee, B. P., Dalsin, J. L., and Burke, S. A., inventors (2008). DOPA-functionized, branched, poly(aklylene oxide) adhesives patent, US Patent Application 20080247984.

86. Ikada, Y. (1997). Tissue adhesives, in *Wound Closure Biomaterials and Devices*, (eds., Chu, C. C., von Fraunhofer, J. A., and Greisler, H. P.), CRC Press, Inc., Boca Raton, Florida, pp. 317–346.

87. Eriksen, J. R., Bech, J. I., Linnemann, D., and Rosenberg, J. (2008). Laparoscopic intraperitoneal mesh fixation with fibrin sealant (Tisseel) vs. titanium tacks: A randomised controlled experimental study in pigs, *Hernia*, **12**, 483–491.

88. Olivier ten Hallers, E. J., Jansen, J. A., Marres, H. A. M., Rakhorst, G., and Verkerke, G. J. (2007). Histological assessment of titanium and polypropylene fiber mesh implantation with and without fibrin tissue glue, *J. Biomed. Mater. Res. Part A.*, **80A**, 372–380.

89. Evans, C. E., Lees, G. C., and Trail, I. A. (1999). Cytotoxicity of cyanoacrylate adhesives to cultured tendon cells, *J. Hand Surg.*, **24B**, 658–661.

90. Fortelny, R. H., Petter-Puchner, A. H., Walder, N., Mittermayr, R., Öhlinger, W., Heinze, A., and Redl, H. (2007). Cyanoacrylate tissue sealant impairs tissue integration of macroporous mesh in experimental hernia repair, *Surg. Endosc.*, **21**, 1781–1785.

91. Harris, J. M. (1992). Introduction to biotechnical and biomedical applications of poly(ethylene glycol), in *Poly(ethylene glycol) Chemistry: Biotechnical and Biomedical Applications* (eds., Harris, J. M.), Plenum Press, New York, pp. 1–14.

92. Zalipsky, S., and Harris, J. M. (1997). Introduction to chemistry and biological applications of poly(ethylene glycol), American Chemical Society, in *Poly(ethylene glycol): Chemistry and Biological Applications* (eds., Harris, J. M., and Zalipsky, S.), Washington, DC, pp. 1–13.

93. Bilic, G., Brubaker, C., Messersmith, P. B., Mallik, A. S., Quinn, T. M., Haller, C., Done, E., Gucciardo, L., Zeisberger, S. M., Zimmermann, R., Deprest, J., and Zisch, A. H. (2010). Injectable candidate sealants for fetal membrane repair: Bonding and toxicity in vitro, *Am. J. Obstet. Gynecol.*, **202**, 85.e1–85.e9.

94. Brubaker, C. E., Kissler, H., Wang, L.-J., Kaufman, D. B., and Messersmith, P. B. (2010). Biological performance of mussel-inspired adhesive in extrahepatic islet transplantation, *Biomaterials*, **31**, 420–427.

95. Haller, C. M., Buerzle, W., Kivelio, A., Perrini, M., Brubaker, C. E., Gubeli, R. J., Mallik, A. S., Weber, W., Messersmith, P. B., Mazza, E., Ochsenbein-Koelble, N., Zimmermann, R., and Ehrbar, M. (2012). Mussel-mimetic tissue adhesive for fetal membrane repair: An ex vivo evaluation, *Acta Biomater.*, **8**, 4365–4370.

96. Haraguchi, K. (2011). Synthesis and properties of soft nanocomposite materials with novel organic/inorganic network structures, *Polym J.*, **43**, 223–241.

97. Sarkar, S., Guibal, E., Quignard, F., and SenGupta, A. (2012). Polymer-supported metals and metal oxide nanoparticles: synthesis, characterization, and applications, *J. Nanoparticle Res.*, **14**, 1–24.

98. Gaharwar, A. K., Dammu, S. A., Canter, J. M., Wu, C. J., and Schmidt, G. (2011). Highly extensible, tough, and elastomeric nanocomposite hydrogels from poly(ethylene glycol) and hydroxyapatite nanoparticles, *Biomacromolecules*, **12**, 1641–1650.

99. Haraguchi, K., Farnworth, R., Ohbayashi, A., and Takehisa, T. (2003). Compositional effects on mechanical properties of nanocomposite hydrogels composed of poly(*N*,*N*-dimethylacrylamide) and clay, *Macromolecules*, **36**, 5732–5741.

100. Wang, T., Liu, D., Lian, C., Zheng, S., Liu, X., and Tong, Z. (2012). Large deformation behavior and effective network chain density of swollen poly(*N*-isopropylacrylamide)-laponite nanocomposite hydrogels, *Soft Matter*, **8**, 774–783.

101. Skelton, S., Bostwick, M., O'Connor, K., Konst, S., Casey, S., and Lee, B. P. (2013). Biomimetic adhesive containing nanocomposite hydrogel with enhanced materials properties, *Soft Matter*, **9**, 3825–3833.

102. Liu, Y., Zhan, H., Skelton, S., and Lee, B. P. (2013). Marine adhesive containing nanocomposite hydrogel with enhanced materials and bioadhesive properties, *Mater. Res. Soc. Proc.*, **1569**, 33–38.

103. Lee, B. P., Dalsin, J. L., Vollenweider, L., Murphy, J. L., Xu, F., Virosco, J., Lew, W., and White, J., inventors (2010). Bioadhesive construct, US20100137902.

104. Lee, B. P., Dalsin, J. L., Vollenweider, L., Murphy, J. L., Xu, F., Virosco, J., Lew, W., and White, J., inventors (2012). Bioadhesive construct with polymer blends. US20120003888.

105. Gelberman, R. H., Boyer, M. I., Brodt, M. D., Winters, S. C., and Silva, M. J. (1999). The effect of gap formation at the repair site on the strength and excursion of intrasynovial flexor tendons: An experimental study on the early stages of tendon-healing in dogs, *J. Bone Joint Surg.*, **81A**, 975–982.

106. Boyer, M. I., Gelberman, R. H., Burns, M. E., Dinopoulos, H., Hofem, R., and Silva, M. J. (2001). Intrasynovial flexor tendon repair an experimental study comparing low and high levels of in vivo force during rehabilitation in canines, *J. Bone Joint Surg.*, **83**, 891–899.

107. Geim, A. K., Dubonos, S. V., Grigorieva, I. V., Novoselov, K. S., Zhukov, A. A., and Shapoval, S. Y. (2003). Microfabricated adhesive mimicking gecko foot-hair, *Nat. Mater.*, **2**, 461–463.

108. Yurdumakan, B., Raravikar, N. R., Ajayan, P. M., and Dhinojwala, A. (2005). Synthetic gecko foot-hairs from multiwalled carbon nanotubes, *Chem. Commun.*, **30**, 3799–3801.

109. Huber, G., Mantz, H., Spolenak, R., Mecke, K., Jacobs, K., Gorb, S. N., and Arzt, E. (2005). Evidence for capillary contributions to gecko adhesion from single spatula nanomechanical measurements, *Proc. Nat. Acad. Sci. U. S. A.*, **102**, 16293–16296.

110. Sun, W., Neuzil, P., Kustandi, T. S., OH, S., and Samper, V. D. (2005). The nature of the gecko lizard adhesive force, *Biophys. J.*, **89**, L14–1L6.

111. Glass, P., Chung, H., Washburn, N. R., and Sitti, M. (2009). Enhanced reversible adhesion of dopamine methacrylamide-coated elastomer microfibrillar structures under wet conditions, *Langmuir*, **25**, 6607–6612.

112. Glass, P., Chung, H., Washburn, N. R., and Sitti, M. (2010). Enhanced wet adhesion and shear of elastomeric micro-fiber arrays with mushroom tip geometry and a photopolymerized p(DMA-co-MEA) tip coating, *Langmuir*, **26**, 17357–17362.

113. Bixler, G. D., and Bhushan, B. (2012). Biofouling: Lessons from nature, *Philos. Trans. R. Soc. A Math. Phys. Eng. Sci.*, **370**, 2381–2417.

114. Zhang, M. Q., Desai, T., and Ferrari, M. (1998). Proteins and cells on PEG immobilized silicon surfaces, *Biomaterials*, **19**, 953–960.

115. Kamath, K. R., Danilich, M. J., Marchant, R. E., and Park, K. (1996). Platelet interactions with plasma-polymerized ethylene oxide and *N*-vinyl-2-pyrrolidone films and linear poly(ethylene oxide) layer, *J. Biomat. Sci.Polym. E.*, **7**, 977–988.

116. Bridgett, M. J., Davies, M. C., and Denyer, S. P. (1992). Control of staphylococcal adhesion to polystyrene surfaces by polymer surface modification with surfactants, *Biomaterials*, **13**, 411–416.

117. Desai, N. P., Hossainy, S. F. A., and Hubbell, J. A. (1992). Surface-immobilized polyethylene oxide for bacterial repellence, *Biomaterials*, **13**, 417–420.

118. Harris, L. G., Tosatti, S., Wieland, M., Textor, M., and Richards, R. G. (2004). *Staphylococcus aureus* adhesion to titanium oxide surfaces coated with non-functionalized and peptide-functionalized poly(L-lysine)-grafted-poly(ethylene glycol) copolymers, *Biomaterials*, **25**, 4135–4148.

119. Pasche, S., De Paul, S. M., Voros, J., Spencer, N. D., and Textor, M. (2003). Poly(L-lysine)-graft-poly(ethylene glycol) assembled monolayers on niobium oxide surfaces: A quantitative study of the influence of polymer interfacial architecture on resistance to protein adsorption by ToF-SIMS and in situ OWLS, *Langmuir*, **19**, 9216–9225.

120. Bain, C. D., and Whitesides, G. M. (1988). Molecular-level control over surface order in self-assembled monolayer films of thiols on gold, *Science*, **240**, 62–63.

121. Zhang, Y., Kohler, N., and Zhang, M. (2002). Surface modification of superparamagnetic magnetite nanoparticles and their intracellular uptake, *Biomaterials*, **22**, 1553–1561.

122. Zhang, F., Kang, E. T., Neoh, K. G., Wang, P., and Tan, K. L. (2001).Surface modification of stainless steel by grafting of poly(ethylene glycol) for reduction in protein adsorption, *Biomaterials*, **22**, 1541–1548.

123. Shen, M. C., Martinson, L., Wagner, M. S., Castner, D. G., Ratner, B. D., and Horbett, T. A. (2002). PEO-like plasma polymerized tetraglyme surface interactions with leukocytes and proteins: in vitro and in vivo studies, *J. Biomat. Sci.-Polym. E*, **13**, 367–390.

124. Schoenfisch, M. H., and Pemberton, J. E. (1998). Air stability of alkanethiol self-assembled monolayers on silver and gold surfaces, *J. Am. Chem. Soc.*, **120**, 4502–4513.

125. Tarlov, M. J., and Newman, J. G. (1992). Static secondary ion mass-spectrometry of self-assembled alkanethiol monolayers on gold, *Langmuir*, **8**, 1398–1405.

126. Statz, A., Finlay, J., Dalsin, J., Callow, M., Callow, J. A., and Messersmith, P. B. (2006). Algal antifouling and fouling-release properties of metal surfaces coated with a polymer inspired by marine mussels, *Biofouling*, **22**, 391–399.

127. Dalsin, J. L., and Messersmith, P. B. (2005). Bioinspired antifouling polymers, *Mater. Today*, **8**, 38–46.

128. Morlat, S., and Gardette, J.-L. (2001). Phototransformation of water-soluble polymers. I: Photo- and thermooxidation of poly(ethylene oxide) in solid state *Polymer*, **42**, 6071–6079.

129. Nelson, C. M., Raghavan, S., Tan, J. L., and Chen, C. S. (2003). Degradation of micropatterned surfaces by cell-dependent and -independent processes, *Langmuir*, **19**, 1493–1499.

130. Statz, A. R., Barron, A. E., and Messersmith, P. B. (2008). Protein, cell and bacterial fouling resistance of polypeptoid-modified surfaces: Effect of side-chain chemistry, *Soft Matter*, **4**, 131–139.

131. Statz, A. R., Meagher, R. J., Barron, A. E., and Messersmith, P. B. (2005). New peptidomimetic polymers for antifouling surfaces, *JACS*, **127**, 7972–7973.

132. Statz, A. R., Park, J. P., Chongsiriwatana, N. P., Barron, A. E., and Messersmith, P. B. (2008). Surface-immobilized antimicrobial peptoids, *Biofouling*, **24**, 439–448.

133. Miller, S. M., Simon, R. J., Ng, S., Zuckermann, R. N., Kerr, J. M., and Moos, W. H. (1995). Comparison of the proteolytic susceptibilities of homologous L-amino acid, D-amino acid, and N-substituted glycine peptide and peptoid oligomers, *Drug Develop. Res.*, **35**, 20–32.

134. Brault, N. D., Gao, C., Xue, H., Piliarik, M., Homola, J., Jiang, S., and Yu, Q. (2010). Ultra-low fouling and functionalizable zwitterionic coatings grafted onto SiO_2 via a biomimetic adhesive group for sensing and detection in complex media, *Biosens. Bioelectron.*, **25**, 2276–2282.

135. Gong, Y.-K., Liu, L.-P., and Messersmith, P. B. (2012). Doubly biomimetic catecholic phosphorylcholine copolymer: A platform strategy for fabricating antifouling surfaces, *Macromol. Biosci.*, **12**, 979–985.

136. Fan, X., Lin, L., and Messersmith, P. B. (2006). Cell fouling resistance of polymer brushes grafted from Ti substrates by surface-initiated polymerization: Effect of ethylene glycol side chain length, *Biomacromolecules*, **7**, 2443–2448.

137. Fan, X., Lin, L., Dalsin, J. L., and Messersmith, P. B. (2005). Biomimetic anchor for surface-initiated polymerization from metal substrates, *JACS*. **127**, 15843–15847.

138. Wang, X., Ye, Q., Gao, T., Liu, J. and Zhou, F. (2011). Self-assembly of catecholic macroinitiator on various substrates and surface-initiated polymerization, *Langmuir*, **28**, 2574–2581.

139. Gao, C., Li, G., Xue, H., Yang, W., Zhang, F., and Jiang, S. (2010). Functionalizable and ultra-low fouling zwitterionic surfaces via adhesive mussel mimetic linkages, *Biomaterials*, **31**, 1486–1492.

140. Ye, Q., Wang, X., Li, S., and Zhou, F. (2010). Surface-initiated ring-opening metathesis polymerization of pentadecafluorooctyl-5-norbornene-2-carboxylate from variable substrates modified with sticky biomimic initiator, *Macromolecules*, **43**, 5554–5560.

141. Waite, J. H., and Qin, X. (2001). Polyphosphoprotein from the adhesive pads of *Mytilus edulis*, *Biochemistry*, **40**, 2887–2893.

142. Lee, H., Rho, J., and Messersmith, P. B. (2009). Facile conjugation of biomolecules onto surfaces via mussel adhesive protein inspired coatings, *Adv. Mater.*, **21**, 431–434.

143. Yang, S. H., Kang, S. M., Lee, K. B., Chung, T. D., Lee, H., and Choi, I. S. (2011). Mussel-inspired encapsulation and functionalization of individual yeast cells, *J. Am. Chem. Soc.*, **133**, 2795–2797.

144. Kang, S. M., Hwang, N. S., Yeom, J., Park, S. Y., Messersmith, P. B., Choi, I. S., Langer, R., Anderson, D. G., and Lee, H. (2012). One-step multipurpose surface functionalization by adhesive catecholamine, *Adv. Funct. Mater.*, **22**, 2949–2955.

145. Ren, Y., Rivera, J. G., He, L., Kulkarni, H., Lee, D. K., and Messersmith, P. B. (2011). Facile, high efficiency immobilization of lipase enzyme on magnetic iron oxide nanoparticles via a biomimetic coating, *BMC Biotechnol.*, **11**, 63.

146. Ham, H. O., Liu, Z., Lau, K. H. A., Lee, H., and Messersmith, P. B. (2011). Facile DNA immobilization on surfaces through a catecholamine polymer, *Angew. Chem. Int. Ed.*, **50**, 732–736.

147. Cheng, C., Li, S., Nie, S., Zhao, W., Yang, H., Sun, S., and Zhao, C. (2012). General and biomimetic approach to biopolymer-functionalized graphene oxide nanosheet through adhesive dopamine, *Biomacromolecules*, **13**, 4236–4246.

148. Fei, B., Qian, B. T., Yang, Z. Y., Wang, R. H., Liu, W. C., Mak, C. L., and Xin, J. H. (2008). Coating carbon nanotubes by spontaneous oxidative polymerization of dopamine, *Carbon*, **46**, 1795–1797.

149. Iqbal, Z., Lai, E. P. C., and Avis, T. J. (2012). Antimicrobial effect of polydopamine coating on *Escherichia coli*, *J. Mater. Chem.*, **22**, 21608.

150. Cheng, C., Li, S., Zhao, W., Wei, Q., Nie, S., Sun, S., and Zhao, C. (2012). The hydrodynamic permeability and surface property of polyethersulfone ultrafiltration membranes with mussel-inspired polydopamine coatings, *J. Membr. Sci.*, **417–418**, 228–236.

151. Guo, L., Liu, Q., Li, G., Shi, J., Liu, J., Wang, T., and Jiang, G. (2012). A mussel-inspired polydopamine coating as a versatile platform for the in situ synthesis of graphene-based nanocomposites, *Nanoscale*, **4**, 5864–5867.

152. Yang, H., Lan, Y., Zhu, W., Li, W., Xu, D., Cui, J., Shen, D., and Li, G. (2012). Polydopamine-coated nanofibrous mats as a versatile platform for producing porous functional membranes, *J. Mater. Chem.*, **22**, 16994.

153. Rai, M., Yadav, A., and Gade, A. (2009). Silver nanoparticles as a new generation of antimicrobials, *Biotechnol. Adv.*, **27**, 76–83.

154. Fullenkamp, D. E., Rivera, J. G., Gong, Y. K., Lau, K. H., He, L., Varshney, R., and Messersmith, P. B. (2012). Mussel-inspired silver-releasing antibacterial hydrogels, *Biomaterials*, **33**, 3783–3791.

155. Son, H. Y., Ryu, J. H., Lee, H., and Nam, Y. S. (2013). Silver-polydopamine hybrid coatings of electrospun poly(vinyl alcohol) nanofibers, *Macromol. Mater. Eng.*, **298**, 547–554.

156. Roe, D., Karandikar, B., Bonn-Savage, N., Gibbins, B., and Roullet, J. B. (2008). Antimicrobial surface functionalization of plastic catheters by silver nanoparticles, *J. Antimicrob. Chemother.*, **61**, 869–876.

157. Sileika, T. S., Kim, H. D., Maniak, P., and Messersmith, P. B. (2011). Antibacterial performance of polydopamine-modified polymer surfaces containing passive and active components, *ACS Appl. Mater. Interfaces*, **3**, 4602–4610.

158. Cui, J., Wang, Y., Postma, A., Hao, J., Hosta-Rigau, L., and Caruso, F. (2010). Monodisperse polymer capsules: Tailoring size, shell thickness, and hydrophobic cargo loading via emulsion templating, *Adv. Funct. Mater.*, **20**, 1625–1631.

159. Ma, Z., Jia, X., Hu, J., Zhang, G., Zhou, F., Liu, Z., and Wang, H. (2013). Dual-responsive capsules with tunable low critical solution temperatures and their loading and release behavior, *Langmuir: ACS J. Surf. Colloids*, **29**, 5631–5637.

160. Zhang, L., Shi, J., Jiang, Z., Jiang, Y., Qiao, S., Li, J., Wang, R., Meng, R., Zhu, Y., and Zheng, Y. (2011). Bioinspired preparation of polydopamine microcapsule for multienzyme system construction, *Green Chem.*, **13**, 300–306.

161. Pan, X.-D., Qina, Z., Yana, Y.-Y., and Sadhukhana, P. (2010). Elastomers with chain-end mussel-mimetic modification for nanocomposites: Strong modifications to reinforcement and viscoelastic properties, *Polymer*, **51**, 3453–3461.

162. Lee, M., Rho, J., Lee, D.-E., Hong, S., Choi, S.-J., Messersmith, P. B., and Lee, H. (2012). Water detoxification by a substrate-bound catecholamine adsorbent, *ChemPlusChem*, **77**, 987–990.

163. Jia, X., Ma, Z. Y., Zhang, G. X., Hu, J. M., Liu, Z. Y., Wang, H. Y., and Zhou, F. (2013). Polydopamine film coated controlled-release multielement compound fertilizer based on mussel-inspired chemistry, *J. Agric. Food Chem.*, **61**, 2919–2924.

164. Fullenkamp, D. E., He, L., Barrett, D. G., Burghardt, W. R., and Messersmith, P. B. (2013). Mussel-inspired histidine-based transient network metal coordination hydrogels, *Macromolecules*, **46**, 1167–1174.

165. Bell, E., and Gosline, J. (1996). Mechanical design of mussel byssus: material yield enhances attachment strength, *J. Exp. Biol.*, **199**, 1005–1017.

166. Tamarin, A., Lewis, P., and Askey, J. (1976). The structure and formation of the byssus attachment plaque in *Mytilus*, *J. Morphol.*, **149**, 199–221.

167. Suresh, S. (2001). Graded materials for resistance to contact deformation and damage, *Science*, **292**, 2447–2451.

Chapter 7

Algal Glue Mimetics

Ronit Bitton

Department of Chemical Engineering, Ben-Gurion University of the Negev, 84105 Beer-Sheva, Israel & Ilse Katz Institute for Nanoscale Science and Technology, Ben Gurion University of the Negev, Beer Sheva 84105, Israel

rbitton@exchange.bgu.ac.il

7.1 Introduction

Soft tissue adhesives are required to perform a variety of functions: stop bleeding, seal leaks, bind tissue and/or facilitate healing [1]. Several types of polymeric tissue adhesives are currently available commercially. Yet, despite considerable developments in the past decade, these adhesives still suffer from serious drawbacks. While synthetic adhesives display low biocompatibility, potential toxicity, and low adherence to wet surfaces, the biological glues are expensive, often show relatively poor mechanical and tissue-bonding properties, and are potentially immunogenic.

One of the most challenging technological aspects that have to be faced while developing new tissue adhesives is to create a material that can glue surfaces under an aqueous environment. The success of synthetic adhesives in a hydrated environment is rather limited. In contrast, many adhesion events in nature occur in highly hydrated conditions and some even under complete submersion

Bioadhesion and Biomimetics: From Nature to Applications
Edited by Havazelet Bianco-Peled and Maya Davidovich-Pinhas

ISBN 978-981-4463-98-0 (Hardcover), 978-981-4463-99-7 (eBook)
www.panstanford.com

in water. Marine sessile organisms, such as mussels, barnacles, tubeworms, starfish and algae, effectively stick to almost any hydrated underwater surface [2]. These natural adhesives, such as "mussel glue," have been proposed to be suitable for medical applications. However, producing a commercially feasible quantity of a natural adhesive from naturally existing raw materials is currently not practical. An alternative method is based on taking a "biomimetic" approach, which entails constructing artificial materials that mimic natural forms. The vast biochemical knowledge on mussels and gecko adhesives accumulated over the past two decades [2] led to the development of several mussel-inspired glues [3–9]. This chapter, however, focuses on another marine creature in which adhesive has been recently explored: the algae.

7.2 Algal Glue

During settlement, algal species produce and secret adhesives that form permanent strong and flexible underwater bonds to a variety of hydrophilic and hydrophobic substrate [10,11]. Typically, the major components of most algal adhesives are acidic, anionic polysaccharides [12,13]. These adhesives can also contain glycoproteins as in the case of *Ulva* and *Palmaria* [11,14] or phenolic compounds as in many red and brown algae [13,15–17]. However, research into their adhesive abilities has been lacking and most of the proposed adhesion mechanisms are mostly speculative.

7.2.1 Green Algae

Up to this point, the most extensive studied algae glue is the one produced by the green algae *Ulva* (formerly known as *Enteromopha*); upon initial selection of the settlement site, the algae spores establish brief reversible adhesion with the substrate. Once permanent settlement is initiated, the spores release an adhesive and undergo morphological changes, including the formation of a cell wall [18]. Electron-dense vesicles in the interior region of the swimming spores are believed to be packaged adhesives. Initial characterization of the adhesive proteins revealed that they were extensively glycosylated (except for glycolsylation, no other post-translational modifications have been detected). After secretion, the glue undergoes rapid swelling followed by a curing process

[19,20]. The curing process of the adhesive is still a mystery. However, it appears that after the secretion of the vesicles, a second set of proteins or agents is packaged and secreted. This second wave has the effect of influencing the curing of the primary adhesive [21]. Despite the growing knowledge on the *Ulva* adhesion mechanism, most efforts are focused on the development of anti-fouling substrates, and to the best of our knowledge, there are no reported studies on biomimetic *Ulva* glues.

7.2.2 Brown Algae

A study of the brown algae *U. pinnatifida* spores, conducted by Petrone et al. [12], identified phosphate groups and monoester-sulfated polysaccharides as key components of the algae adhesive; in fact, algae grown in a sulfate-deprived environment did not have the ability to settle. This knowledge has not been utilized for the development of new synthetic adhesives.

Another extensively studied family of algae glues is formed by the brown algae *Fucus*. Vreeland and Epstein [13] postulated that initial substratum adhesion by zygotes of the brown alga *Fucus gardenri* involves the secretion of polyphenols. Later, after germination, the phenolic polymer secretion was localized at the site of attachment indicating that phenolic polymers and oxidases might play a key role in the algae adhesion. They proposed a "fiber-phenolic-catalyst" mechanism (Fig. 7.1) whereby the secretion of phenolic compounds coupled with peroxidase results in the cross-linking of cell-wall polysaccharides [22].

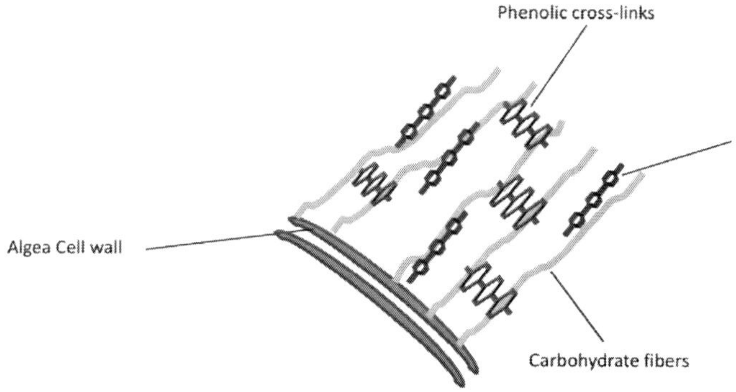

Figure 7.1 A scheme describing the hypothesized structure of polyphenol-carbohydrate glue. Adapted from ref. 22.

NMR study of polyphenols extracted from *Fucus vesiculus* showed that brown algal polyphenol are composed of phloroglucinol units linked by carbon-carbon and ether bonds [23]. Berglin et al. [24] studied the enzymatic cross-linking of a phenolic polymer extracted from *Fucus serratus* using the quartz crystal microbalance with dissipation monitoring methodology (QCM-D). Their results show that addition of vanadium-dependent haloperoxidase (BPO) enzyme, KBr and H_2O_2 to the phenolic polymer caused a decrease in dissipation, indicating that a cross-linking process may have occurred. Although all four components were proved to be necessary, the cross-linking mechanism remained unclear. Two feasible mechanisms were suggested based on the QCM-D, Surface Plasmon Resonance (SPR) and UV Spectroscopy results: (1) Bromation of the phenols, followed by the oxidation and combination and finally the rearrangement of HBr or (2) Enzymatic generation of oxidized bromine species, which farther reacts with the nucleophilic phloroglucinol repeating unit. However, as suggested in a patent by Vreeland and Epstein [25], alginate and calcium ions should be also used in the adhesive formulations. The adhesion properties of various reconstructed *Fucus* glue formulations, clarifying what is the role of each of the glue's components were investigated by Bitton et al. [26]. The influence of the polyphenol oxidation conditions on the adhesion kinetics and the glue's adhesive strength were also reported. The differences in the adhesion strengths of oxidized and non-oxidized *fucus* glues are particularly noticeable when the adherands were glass (hydrophilic) and Teflon (hydrophobic). As suspected the oxidized glue adhered much better to the hydrophobic surface [27].

Given that *Fucus glue* is a multicomponent polymeric material that can be arranged in a number of spatial arrangements, its structure-property relationships are of particular interest. The structural features of materials extracted from the brown alga *Fucus serratus*, were investigated in order to elucidate the structure-property relationships of the *Fucus serratus* glue, toward a better understanding of the adhesion mechanisms [28]. Using a combination of small angel scattering (SAXS), static light scattering (SLS), and transmission electron microscopy at cryogenic temperature (cryo-TEM), it was discovered that the phenolic polymer forms mostly flexible chain-like objects (Fig. 7.2). The SAXS data fitted nicely to a model of semi-flexible chain with excluded volume [29] with the best fit parameters of the chain contour length (L) = 21000 ± 500 Å, persistence length (l_p) = 120 ± 5 Å and cross section radius (R_{cs}) = 18 ± 2 Å. These findings suggested that the scattering is not due to a single macromolecule, but rather arises from either associated or branched polymeric chains. Indeed, in the cryo-TEM micrographs in Fig. 7.2, one can see chain-like objects having a diameter of about 40 Å. Oxidation of the polyphenol with BPO, KI, and H_2O_2 (verified by NMR) does not

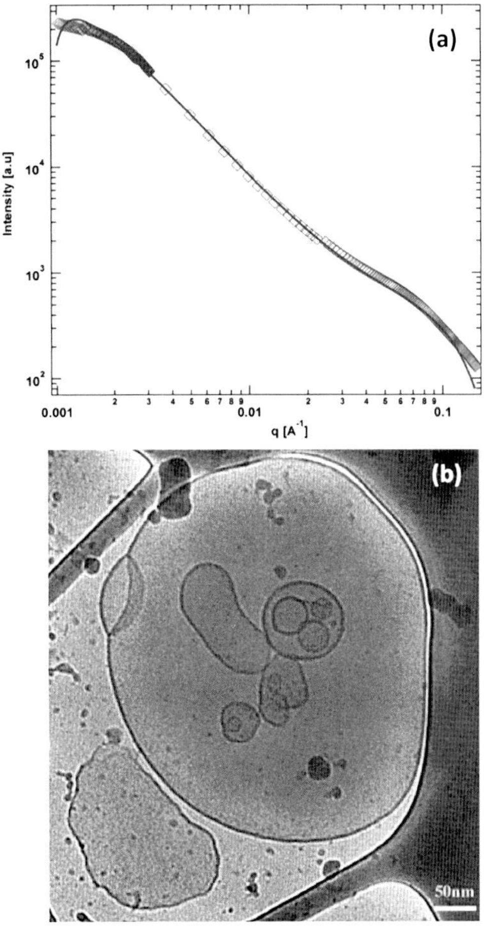

Figure 7.2 (a) SLS and SAXS combined curve for 5 mg/mL native *Fucus serratus* polyphenol in aqueous solution, (b) Cryo-TEM image of 5 mg/mL native *Fucus serratus* polyphenol in aqueous solution.

alter the overall structure of the chain-like objects, causing only a decrease in the cross section radius. Once alginate and calcium ions were added, glue was formed; SAXS data (Fig. 7.3) shows that the overall structure of the glue seems to resemble that of the alginate gel (i.e., a 3D network). Based on these findings, the authors speculated that the polyphenol (either oxidized or natural) is encapsulated by the alginate gel network, as illustrated schematically in Fig. 7.3. Presumably, this arrangement allows the polyphenol to form contact points with the surface, without altering the cohesive strength of the alginate gel.

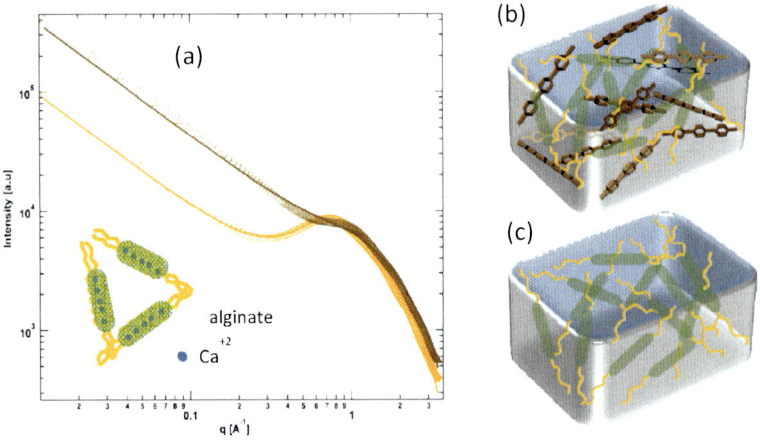

Figure 7.3 (a) SAXS curve from alginate gel (○) and a mixture of oxidized Fucus serratus polyphenol and alginate gel (Δ). The solid lines represent the fit results of the "broken rod linked by a flexible chain" model. Inset: schematic representation of the "broken rod" model; (b) schematic illustration of the spatial arrangement of the "fucus glue"; (c) schematic illustration of the spatial arrangement of alginate gel.

7.3 Biomimetic Algal Glue

Following the findings of the above-mentioned study of the *Fucus* glue's nanostructure, a biomimetic analogue was developed by replacing the natural *Fucus* extracted polymers with new synthetic elements [27]. Aimed at finding glue that will have similar capabilities to the natural glue, yet its production will not rely on the complex and time-consuming process of extracting natural polyphenols from algae; the polyphenol has been replaced by its monomeric unit, phloroglucinol (Fig. 7.4). Glue formulations composed of phloroglucinol, alginate and calcium ions were shown to be capable of adhering to a variety of hydrophilic and hydrophobic surfaces. A comprehensive study of this biomimetic glue's adhesion properties, nanostructure and rheological behavior was performed. The adhesive strength of the biomimetic glue was shown to be comparable to that of the alga-born one (Fig. 7.4). As with the algal-born glue, the oxidized formula adhered better to the Teflon while the non-oxidized formula adhered better to the glass. Although the non-oxidized glue's superiority in adhering hydrophilic surfaces is not as clear as it was in the Fucus glue, the non-oxidized formula seems to be better suited for use on hydrated tissues.

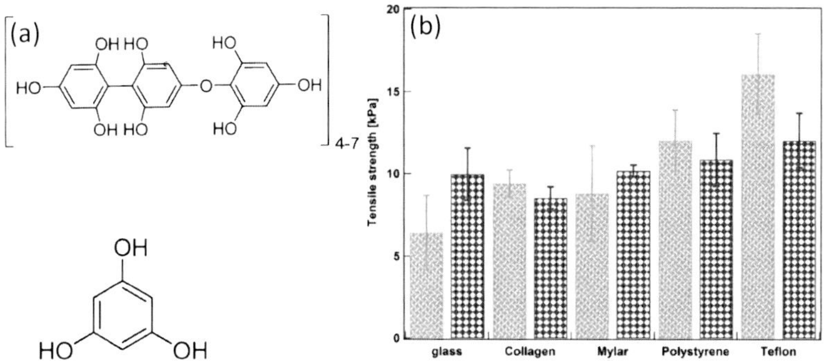

Figure 7.4 (a) Chemical structures of Fucus polyphenol and Phloroglucinol. (b) Tensile strength of Fucus glue (yellow) and biomimetic glue (purple) to various substrates.

As can be seen from SAXS experiments, the overall structure of the glue resembles that of alginate gel. Yet, since the scattering from phloroglucinol by itself was very weak and a scattering curve could not be obtained, the differences between the two curves (Fig. 7.5), can be attributed to interactions between the alginate and the phloroglucinol. Rheology measurements showed that the presence of phloroglucinol lowers the amount of Ca^{+2} ions required for sol-gel transition of alginate, also indicating interactions between the alginate and the phloroglucinol. It was suggested that the cohesive strength is mainly due to alginate gelation, while the adhesion mechanism relies both on mechanical interlocking and weak interactions such as hydrogen bonds and hydrophobic interactions between the phloroglucinol and the surface. Characterization of the adherence to tissues showed that adhesion was directly related to the mechanical strength of the cross-linked alginate. Therefore, the adhesion strength can be altered by changing the source of the calcium cross-linker, the alginate G-content or the molecular weight of the alginate [30].

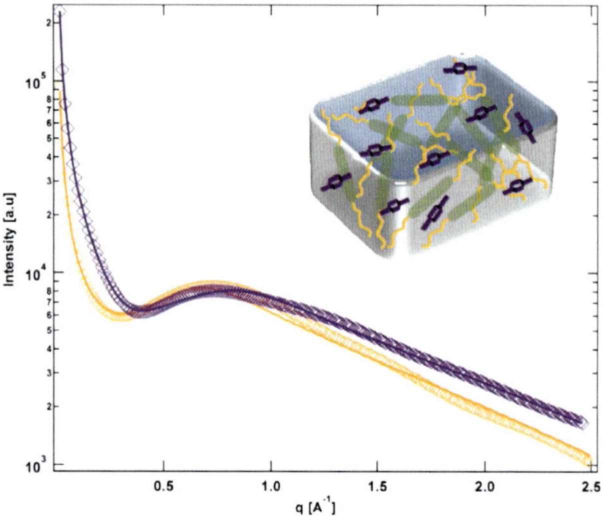

Figure 7.5 SAXS curve from alginate gel (○) and from a biomimetic "phloroglucinol glue" (□). The solid lines represent the fit results of the "broken rod linked by a flexible chain" model. Inset: schematic illustration of the spatial arrangement of the biomimetic "phloroglucinol glue."

The feasibility of the biomimetic formulation to perform as a tissue adhesive was also investigated [30]. The cytotoxicity of the glue was assessed by monitoring the viability of (i) live cell culture (fibroblasts) with soluble compounds in the culture medium and (ii) live cell culture (fibroblasts) around a hardened glue drop in the center of the culture dish. Both tests show that the key ingredients of a glue containing the natural (i.e., non-oxidized) phloroglucinol do not interact adversely with the cells (Fig. 7.6). The phloroglucinol adhesive bound porcine tissues together with adhesive strength of 17–25 kPa, suggesting that it offers appropriate mechanical properties for application as soft tissue adhesive.

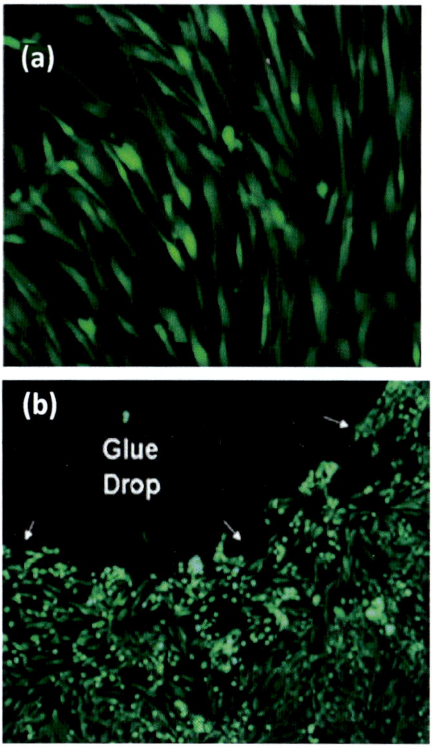

Figure 7.6 (a) Fibroblast cytotoxicity using biomimetic glue, (b) Glue drop assay showing live fibroblasts by fluorescent viability assay, around the perimeter of the hardened glue drop (arrows).

Rozen et al. [31] expanded the study of the above-mentioned bioinspired glue; aiming to lay the basis for the development of a new family of phenolic-based bioinspired sealants. In order to do so, they (1) developed a simple yet reliable apparatus for in vitro characterization of the sealing ability of these adhesives (Fig. 7.7) and (2) replaced the phloroglucinol with other soluble polyphenol in order to expand range of available formulations and possibly to improve the glue's adhesion/sealing capabilities. Burst tests measurements demonstrated the significant divergences between the polyphenols showing that enhanced surface hydrophobicity significantly reduced the sealant performance (Fig. 7.7), suggesting that hydrogen bonding, and not hydrophobic interactions as initially assumed, plays a major role in the adhesion mechanism;

strengthening the hypothesis that phenols interfere (for good or bad) with the sealant–substrate interface.

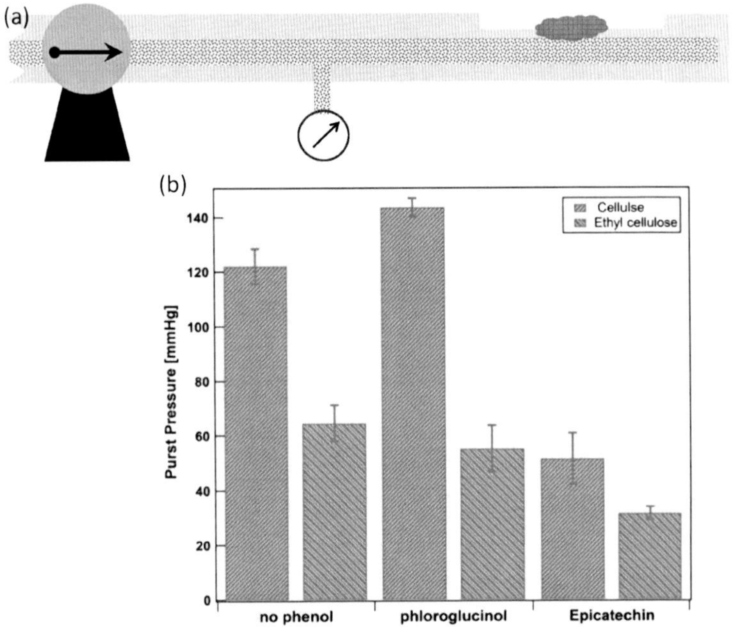

Figure 7.7 (a) Schematic drawing of the burst tester. (b) Burst tests results for sealant with PHG or epicatechin on cellulose and ethyl cellulose substrates.

Based on the above-mentioned findings, a commercial product, Seal-V™, a vascular sealant for use in a broad range of vascular surgical procedures was developed, showing the potential of algae glue as inspiration for novel tissue adhesives. Yet, the performance of the biomimetic glue is still inferior compared to that of the natural algea glue. Further research of algal adhesives has the potential to lead to the development of novel biomimetic, protein free, strong, safe, biocompatibale and biodegradable tissue adhesives.

Acknowledgment

R. B. gratefully acknowledges the support of The Joseph and May Winston Foundation Career Development Chair in Chemical Engineering.

References

1. McDermott, M. K., et al. (2004). Mechanical properties of biomimetic tissue adhesive based on the microbial transglutaminase-catalyzed crosslinking of gelatin. *Biomacromolecules*, **5**(4), 1270–1279.
2. Waite, J. H. (1987). Nature's underwater adhesive specialist. *Int. J. Adhes. Adhes.*, **7**(1), 9–14.
3. Yu, M., and Deming, T. J. (1998). Synthetic polypeptide mimics of marine adhesives. *Macromolecules*, **31**(15), 4739–4745.
4. Catron, N. D., Lee, H., and Messersmith, P. B. (2006). Enhancement of poly(ethylene glycol) mucoadsorption by biomimetic end group functionalization. *Biointerphases*, **1**(4), 134–141.
5. He, L., et al. (2011). pH responsive self-healing hydrogels formed by boronate-catechol complexation. *Chem. Commun.*, **47**(26), 7497–7499.
6. Huang, K., Lee, B., and Messersmith, P. B. (2001). Synthesis and characterization of self-assembling block copolymers containing adhesive moieties. *Polymer Preprints (American Chemical Society, Division of Polymer Chemistry)*, **42**(2), 147–148.
7. Lee, B. P., Dalsin, J. L., and Messersmith, P. B. (2002). Synthesis and gelation of DOPA-modified poly(ethylene glycol) hydrogels. *Biomacromolecules*, **3**(5), 1038–1047.
8. Lee, B. P., et al. (2004). Synthesis of 3,4-dihydroxyphenylalanine (DOPA) containing monomers and their co-polymerization with PEG-diacrylate to form hydrogels. *J. Biomater. Sci., Polymer Ed.*, **15**(4), 449–464.
9. Messersmith, P. B. (2006). Mussel adhesive protein mimetics: polymer-peptide bioconjugates for tissue adhesion and antifouling surfaces. *PMSE Preprints*, **94**, 129.
10. Duarte, A. P., et al. (2012). Surgical adhesives: Systematic review of the main types and development forecast. *Prog. Polymer Sci.*, **37**(8), 1031–1050.
11. Petrone, L., (2013). Molecular surface chemistry in marine bioadhesion. *Adv. Colloid Interface Sci.*, **195**, 1–18.
12. Petrone, L., et al. (2011). In situ ATR-IR spectroscopic and electron microscopic analyses of settlement secretions of Undaria pinnatifida kelp spores. *J. R. Soc. Interface*, **8**(56), 410–422.
13. Vreeland, V., Waite, J. H., and Epstein, L. (1998). Polyphenols and oxidases in substratum adhesion by marine algae and mussels. *J. Phycol.*, **34**(1), 1–8.

14. Iken, K., et al. (2003). A new antifouling bioassay monitoring brown algal spore swimming behaviour in the presence of echinoderm extracts. *Biofouling*, **19**(5), 327–334.
15. Colin, C., et al. (2005). Vanadium-dependent iodoperoxidases in *Laminaria digitata*, a novel biochemical function diverging from brown algal bromoperoxidases. *J. Biol. Inorgan. Chem.*, **10**(2), 156–166.
16. Colin, C., et al. (2003). The brown algal kelp *Laminaria digitata* features distinct bromoperoxidase and iodoperoxidase activities. *J. Biol. Chem.*, **278**(26), 23545–23552.
17. Potin, P., et al. (2005). Integration of molecular biology and chemical approaches for understanding cross-linking mechanisms in brown algal adhesion. *Phycologia*, **44**(4), 82–83.
18. Callow, J. A., et al. (2000). Cellular and molecular approaches to understanding primary adhesion in Enteromorpha: An overview. *Biofouling*, **16**(2–4), 141–150.
19. Callow, J. A., et al. (2003). Use of environmental scanning electron microscopy to image the spore adhesive of the marine alga Enteromorpha in its natural hydrated state. *Colloids Surfaces B-Biointerfaces*, **27**(4), 315–321.
20. Heydt, M., et al. (2012). Settlement behavior of zoospores of ulva linza during surface selection studied by digital holographic microscopy. *Biointerphases*, **7**(1–4), 33. doi: 10.1007/s13758-012-0033-y.
21. Lucas, J. M., Vaccaro, E., and Waite, J. H. (2003). Extra-organismic adhesive proteins. *Biopolymers*, **8**, 359–382.
22. Vreeland, V., and Epstein, L. (1996). Analysis of plant-substratum adhesives. *Modern Methods of Plant Analysis*, **17** (Plant Cell Wall Analysis), 95–116.
23. McInnes, A. G., et al. (1985). Physodes and the phenolic compounds of brown algae. Part 5. The high molecular weight polyphloroglucinols of the marine brown alga Fucus vesiculosus L. Proton and carbon-13 nuclear magnetic resonance spectroscopy. *Can. J. Chem.*, **63**(2), 304–13.
24. Berglin, M., et al. (2004). Enzymatic cross-linking of a phenolic polymer extracted from the marine alga fucus serratus. *Biomacromolecules*, **5**(6), 2376–2383.
25. Vreeland, V., and Grotkopp, E. (1996). *Aqueous water-resistant phloroglucinol-type adhesives and glues derived and extracted from*

algae and activated by oxidizing agents (University of California, USA). Application: US. US. p. 20 pp.

26. Bitton, R., et al. (2007). The influence of halide-mediated oxiodation on algae-born adhesives. *Macromol. Biosci.*, **7**(12), 1280–1289.
27. Bitton, R., and Bianco-Peled, H. (2008). Novel biomimetic adhesives based on algae glue. *Macromol. Biosci.*, **8**(5), 393–400.
28. Bitton, R., et al. (2006). Structure of algal-born phenolic polymeric adhesives. *Macromol. Biosci.*, **6**(9), 737–746.
29. Pedersen, J. S., and Schurtenberger, P. (1996). Scattering functions of semiflexible polymers with and without excluded volume effects. *Macromolecules*, **29**(23), 7602–7612.
30. Bitton, R., et al. (2009). Phloroglucinol-based biomimetic adhesives for medical applications. *Acta Biomater.*, **5**(5), 1582–1587.
31. Y. Rozen and H. Bianco-Peled, Studies of Phenol-Based Bioinspired Sealants, *The Journal of Adhesion*, **90**, 1–14, 2014.

Chapter 8

Bio-Inspired Surfaces with Directional Adhesion

Luciano Afferrante

Dipartimento di Meccanica, Matematica e Management,
Politecnico di Bari, Viale Japigia, 182, Bari, 70126, Italy

luciano.afferrante@poliba.it

This chapter reviews the direction-dependent adhesion properties of microstructured bio-inspired surfaces, consisting of parallel elastic wall-like structures topped by a thin film. Such a system shows high adhesion strength when loaded with a detaching moment acting perpendicularly to the wall-like structures (the critical moment for detachment is higher, 15 times or more, than the one occurring in a flat control sample). When the system is, instead, loaded with a moment acting in any other direction, the critical moment for detachment is less than the one of a flat control sample.

8.1 Introduction

Adhesion in insects, spiders, and lizards must satisfy two requirements. The adherence to a substrate must be strong when

Bioadhesion and Biomimetics: From Nature to Applications
Edited by Havazelet Bianco-Peled and Maya Davidovich-Pinhas

ISBN 978-981-4463-98-0 (Hardcover), 978-981-4463-99-7 (eBook)
www.panstanford.com

the animal wants to secure itself and, at the same time, it must be able to move quickly when necessary. Such abilities are primarily controlled by mechanics rather than surface chemistry. Hence, the geometry and the material properties play a fundamental role in enhancing the adhesion.

Early micropatterned surfaces, inspired by the observation of natural systems, are of increasing scientific interest because of their enhanced superhydrorepellent [1,2] or adhesive [3,4] properties.

For example, the Tokay Gecko has extraordinary adhesion properties due to the hierarchical structure of their toe pads [5–9]. Geckos are capable of attaching and detaching their adhesive toes in few milliseconds [10] while running, a challenge no conventional adhesive is capable of meeting. The properties of gecko setae are "(i) directional, (ii) attaches strongly with minimal preload, (iii) detaches quickly and easily, (iv) sticks to nearly every material, (v) does not stay dirty or (vi) self-adhere, and (vii) is non-sticky by default" [5].

Some gecko species show adhesion strength capabilities of the order of 100 kPa [5]. In geckos, the oriented fibers are made of a stiff biomaterial (β-keratin) with Young's modulus 4 GPa [11] and have diameters from 0.2 to 5 μm [12]. The structure and material properties allow the fibers to individually bend and adapt to a wide variety of surface roughnesses and also to return to their original shape after release from the surface.

In the past years, many scientists are focused their research on the understanding of the principles underlying the gecko adhesive capacities. Also, many attempts to conceive and to fabricate artificial surfaces with adhesive properties similar to the gecko toe pads have been done [3,13–21]. In particular, switchability in adhesion has only recently been explored by Kamperman and Synytska [22]. Examples include systems that respond to external stimuli such as temperature [23–24], magnetic field [25], mechanical stretching [26] and pneumatic pressure [27]. In Paretkar et al. [28] research the adhesion dependence on preload was established for PDMS fibrils, showing that, due to buckling instability of the fibrils, the pull-off strength is high at low preloads, whereas loss in adhesion occurs at high preloads.

Directional adhesives can also be mimicked synthetically in polymer microfiber structures [29–31], for applications in wall-climbing robots [21] or in transfer printing with continuous roll-to-roll applications [32]. In fact, angled microfibers with saucer-shaped tip endings enable high adhesion strength in the gripping direction by combined vertical and lateral loading; they have significantly less strength in the release direction for power efficient and fast detachment.

Fabrication of gecko-like synthetic structures would enable the production of long-lifetime reusable fibrillar adhesives, which could find application in many field such as robotics, biomedical devices, manufacturing, and consumer products.

It is theoretically demonstrated that fibrillar structures may increase the adhesive properties with respect to flat surfaces, even though this behavior is not always reproduced in experimental tests [33–34]. However, Jagota and Bennison [35] have founded that the work required to separate a fibrillar surface is much higher than the work of adhesion calculated for a perfectly bonded non-fibrillar interface. Furthermore, Persson [36] has demonstrated that fibrillar interface have advantages in terms of adhesion with respect to rough surfaces. Significantly improved adhesive capacities are obtained when a terminal plate is added at the end of pillars microstructured on a flat surface, as experimentally observed in Glassmaker et al. [37], where a new architecture, constituted by protruding fibrils topped by a thin film and inspired by the biological setal system found in the insect *Tettigonia viridissima* [38–40], has been proposed.

Examples exist in Nature, e.g., the attachment pad of the males of some beetle species from the family Chrysomelidae, which do not present a hierarchical geometry as gecko does but are simply surfaces covered with mushroom-shaped microstructures [41–43]. In such cases, the shape of the terminal plate is crucial for the achievement of high adhesive strength values of the bio-inspired artificial surfaces [44–45]. Figure 8.1 summarizes the possible debonding modes for simple cylindrical pillars adhering to a rigid substrate: (i) crack propagation from the pillar edge, (ii) propagation of interfacial defects from the inner side, and (iii) decohesion due to the achievement of theoretical contact strength.

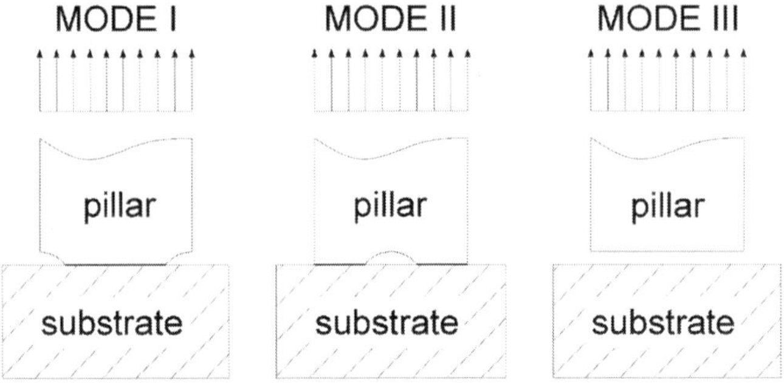

Figure 8.1 Possible detachment modes for a cylindrical pillar rearranged from Ref. [30].

Usually, mode I is more critical, but in mushroom-shaped pillars with a conveniently designed geometry of the terminal plate, crack propagation from the external perimeter of the plate cannot occur since the plate suppresses the stress singularity at the external perimeter, thus resulting in a considerable improvement of adhesive performance.

Also, in Varenberg and Gorb's [46] research, the mushroom-shaped pillars' behavior has been studied during different stages of attachment–loading–detachment cycle showing that the mushroom-shaped geometry of contact elements promotes fast and simple generation of reliable adhesion.

A preferred adhesion direction would make an adhesive particularly well suited to climbing vertical surfaces where the body weight of an animal or robot induces a large downward shear loading on the contacts, and where feet must be detached without disturbing the grip of the remaining legs.

There are many engineering applications (e.g., robot locomotion and object manipulation), where adhesion force acts in a preferential direction, and no adhesion is required along the other direction to facilitate system detachment.

To this purpose, some scientists have created surface microstructures producing high level of shear and adhesion [20–21,47]. Most recently, Parness et al. [48] have proposed a microstructure that allows the surface to act as a controllable adhesive and where the adhesion is controlled by varying the applied shear force. This structure is composed by microfabricated wedge-shaped adhesive array and shows increasing adhesion when shear force is applied in a preferred direction, allowing the amount of adhesion to be controlled. When the shear force is removed, the adhesive can be detached with zero pull-off force.

Efficient climbing relies on this behavior. Materials that require a substantial preload to achieve adhesion and a similarly large force to achieve detachment are not recommended. Robotic climbing with such materials is inefficient and also unreliable. Not surprisingly low detachment forces are also a critical feature of the rapid attachment/detachment cycles observed in the gecko [8].

Despite this promising performance, such type of structures show significantly reduced adhesion capacities for large patches because they fail to make close contact with the climbing surface. In this respect, the gecko employs a hierarchical structure that ensures intimate contact between the nanoscale spatulae and the climbing surface.

Moreover, another challenge in applying synthetic adhesive to robots or other macroscale applications is the prevention of buckling and peeling.

Buckling instability occurs after a critical buckling force P_{cr} is reached. High values of P_{cr} can be obtained using a stiff material (in geckos, the fibers are made of β-keratin). However, the use of a stiff material makes challenging to fabricate a hierarchical structure, similar to the gecko's, which would allow directional adhesives to work at the macroscale.

Peeling occurs when a patch begins to peel at one corner or edge, so the stress concentration can quickly propagate across the entire patch, resulting in a complete adhesive failure.

The architecture proposed in Glassmaker et al. [37] and constituted by protruding fibrils topped by a thin plate have some crack trapping ability, so as to enhance the adhesion capacities. Such type of structure has also the advantage of not suffering buckling, thanks to the presence of the terminal plate connecting the fibrils. However, directionality in adhesion is completely lost.

A step forward in such sense is done in Afferrante and Carbone [49] where a new design of biomimetic microstructured surface, which exhibits controlled strongly direction-dependent adhesion properties, is proposed. The microsystem consists of parallel elastic wall-like structures covered by a thin layer, as shown in Fig. 10.2.

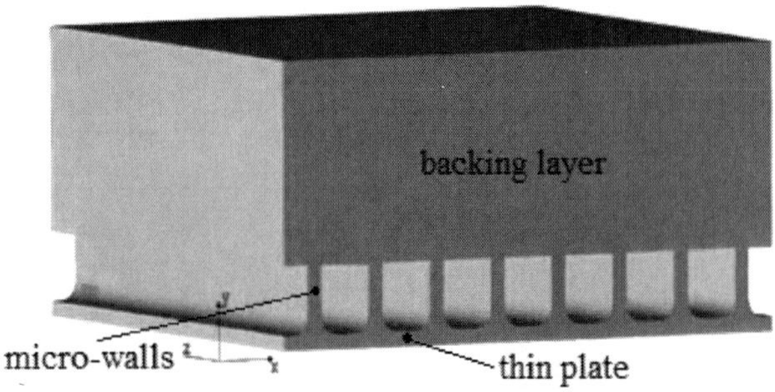

Figure 8.2 A microstructured surface with parallel elastic micro-walls connected with a terminal thin plate. In this chapter, we review this work giving the mathematical basis to explain why similar structures present enhanced adhesion capacities.

Such a type of structure shows strongly enhanced adhesion when loaded with a moment acting in the x–y plane, i.e., when detachment occurs perpendicularly to the walls. For moment acting in any other plane, no increase of adhesion is detected. Such system mimics the direction-dependent adhesion of some biological systems, and may be successfully exploited in some engineering devices where a preferential direction of adhesion is requested so to make possible manipulation and locomotion.

8.2 Energy Release Rate

With reference to Fig. 8.2, the thickness of the backing layer is assumed to be significantly larger than the high of the walls, so as to correctly reproduce the dimensions of the microstructure with respect to the supporting layer of a real geometry. The thin film topping the walls' ends makes contact against a rigid flat surface. Also, the walls, the layer and the thin film are assumed to be linear elastic with Young's modulus E = 3 MPa and Poisson's ratio ν = 0.5 (to simulate a material nearly incompressible).

8.2.1 Detachment Occurring Perpendicularly to the Microwalls

In such case, the crack moves perpendicularly to the microwalls and the system can be simplified, within acceptable tolerances, into a plane strain 2D structure, as shown in Fig. 8.3.

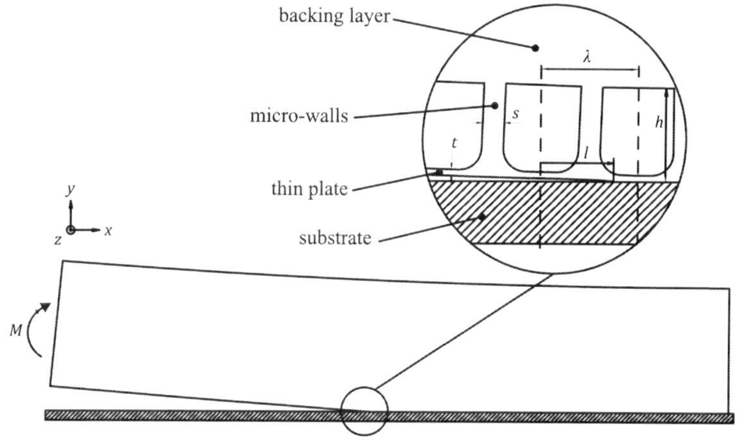

Figure 8.3 A cross section in the x–y plane of the parallel microstructured walls, connected with a terminal film.

The right end of the system is assumed to be sufficiently far from the crack tip may neglect border effects. An external bending moment M is applied to the left edge. For the crack propagation to occur, the energy release rate G at the crack tip (i.e., the elastic strain energy released locally from the material per unit length l) needs to be larger than the thermodynamic work of adhesion γ.

The energy release rate G at the crack tip is evaluated according to the thermodynamic approach presented in Maugis [50].

In particular, denoting with b the transversal width of the system (along the z-axis) and with l the crack length, the strain energy rate G writes as

$$G(l) = -\frac{1}{b}\left(\frac{\partial U_{\mathrm{el}}}{\partial l} + \frac{\partial U_{\mathrm{P}}}{\partial l}\right)\Bigg|_{M=\mathrm{const}}, \tag{8.1}$$

where

$$U_{\mathrm{el}} = \frac{1}{2}M\theta; \qquad U_{\mathrm{P}} = -M\theta \tag{8.2}$$

are, respectively, the elastic energy stored in the system and the potential energy associated with the applied bending moment M (being θ the rotation angle of the left edge of the system where the moment M is applied).

Now, if we define the stiffness of the system, which is a function of the extension of the crack, as $K(l)$, we can write

$$M = K(l)\theta \tag{8.3}$$

and hence

$$G(l) = -\frac{1}{2b}\frac{M^2}{K(l)^2}\frac{\partial K(l)}{\partial l} = \frac{M^2}{2b}\frac{\partial C(l)}{\partial l}, \tag{8.4}$$

where $C = 1/K$ is the system compliance, which changes as the crack advances a unit length.

When the crack moves perpendicularly to the walls, i.e., along the x-axis, the periodic geometry of the system will force the energy release rate G to vary periodically with the spatial period λ. The change in stored elastic energy that occurs when the crack advances a spatial period λ, can be written as $\Delta U_{\mathrm{el}} = -1/2\,\Delta U_{\mathrm{p}}$. Then the average value of $G(l)$ is

$$G_{\mathrm{ave}} = \frac{1}{\lambda}\int_{\lambda} G(l)dl = -\frac{1}{b\lambda}(\Delta U_{\mathrm{el}} + \Delta U_{\mathrm{P}}) = \frac{\Delta U_{\mathrm{el}}}{b\lambda} \tag{8.5}$$

The spontaneous evolution of the system will occur when $G > \gamma$. To calculate $G(l)$, a finite element analysis can be performed. The sticking friction of the zone adhering to the flat rigid substrate, experimentally observed in Varenberg and Gorb [46], can be taken into account by properly constraining the nodes of the adhering surface.

Figure 8.4 shows the variation of the energy release rate G available to the crack tip, normalized with respect to that of a flat control structure G_{flat}, as a function of the dimensionless crack length l/λ, being λ the microwalls spacing. In particular, the effect of the geometrical parameters (the thickness t of the thin film, the spacing λ between the walls, the thickness s and height h of the walls) on the adhesive properties of the system is investigated.

Notice, as a result of the periodicity of the system, G/G_{flat} is a periodic function of l/λ. Also, the variation of G with the crack length l is positive ($dG/dl > 0$) when the crack extends under a wall (elastic energy is released from the strip), and it is negative

($dG/dl < 0$) when the crack tip is between the walls (elastic energy is absorbed into the strip).

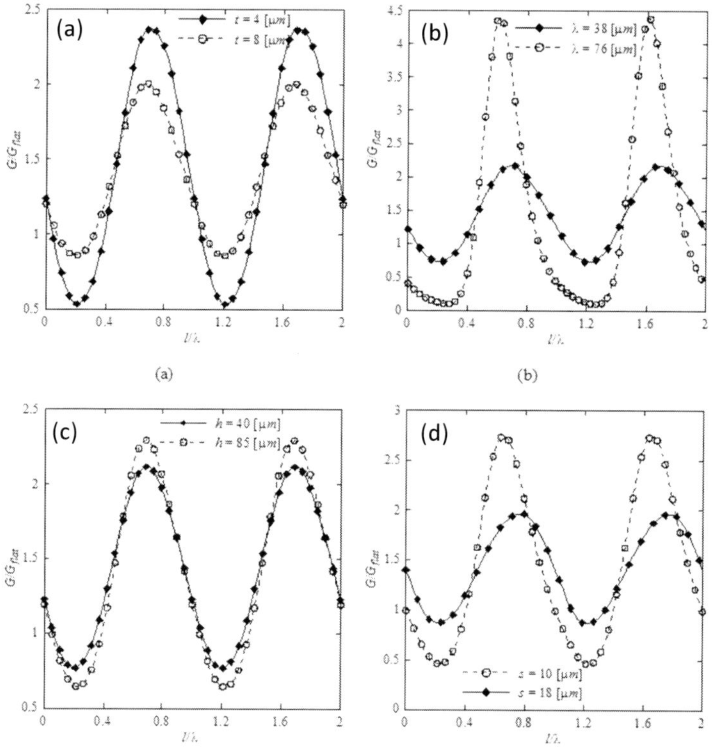

Figure 8.4 Variation of the energy release rate of the system G, normalized with respect to that of a flat control sample G/G_{flat}, with the dimensionless crack length l/λ, for different: (a) film thicknesses t (s = 4 μm; h = 85 μm; λ = 95 μm); (b) wall spacing λ (t = 2 μm; s = 4 μm; h = 85 μm); (c) wall heights h (t = 2 μm; s = 4 μm; λ = 95 μm); (d) wall thicknesses s (t = 2 μm; h = 85 μm; λ = 95 μm). Material constants: E = 3 MPa; ν = 0.5. Applied moment: $M = 9.5*10^{-2}$ N m/m.

This behavior has been qualitatively observed in experimental tests [33–34]. Notice the maximum value of the energy release rate occurs when the crack tip has been passing a wall (we are assuming the crack extends from left to right).

Just to clarify things, consider the two cases corresponding to the film thicknesses t = 4 μm and t = 8 μm and assume $\gamma/G_{\text{flat}} = 0.75$.

For t = 4 μm, crack propagation is stable being $dG/dl < 0$. In fact, the crack propagates up to the value of l/λ for which $G = \gamma$. Further propagation is not possible because it would result in values of G less than the work of adhesion.

For t = 8 μm, the energy release rate is always larger than the work of adhesion γ and unstable crack growth occurs, resulting in a complete detachment of the system.

Therefore, the crack trapping behavior, above described, results in an enhancement of fracture toughness and the best performance would seem to be obtained by minimizing the thickness of the terminal film. With similar argumentations, from Figs. 8.4b–d, one would may deduce that the adhesive properties can be optimized increasing the walls spacing λ and reducing the walls thickness s, whereas the effect of the walls height h is almost negligible.

Moreover, as t becomes vanishingly small, the minimum $G_{\min}$ of the energy release rate goes to zero, so that only if the input of energy is infinite (or equivalently if the work of adhesion is zero) the system can be detached. In reality, it is not necessary an infinite input of energy to detach the strip, because for a large input of energy, large tensile stresses develop under the walls, and failure occurs by mode II or mode III, as depicted in Fig. 8.1.

8.2.2 Detachment Occurring Obliquely to the Microwalls

When the crack propagation is not perpendicular to the walls, the system compliance almost linearly increases with the crack length. As a result, the energy release rate $G(l)$ at the crack tip practically remains constant during the crack propagation and the crack trapping phenomenon cannot occur.

8.3 The Critical Moment for Detachment

8.3.1 Detachment Occurring Perpendicularly to the Microwalls

For a crack propagating perpendicularly to the wall-like structures, assuming that the detachment follows the mode I debonding mechanism [44], i.e., it is governed by the propagation of the crack from the edge of the system, the moment M_I necessary to activate the process can be calculated by enforcing the condition $G_{min} = \gamma$, i.e., using Eq. (8.4):

$$M_I = \left(\frac{2K(l)^2 \gamma b}{\partial K(l)/\partial l|_{min}} \right)^{1/2} \quad (8.6)$$

However, we know that other debonding mechanisms are possible [44], because for a large input of energy, large tensile stresses develop under the walls, and even if failure does not occur by crack propagation (mode I), the interface must fail because either the stresses exceed the interfacial strength (mode III) or a debonding mechanism with a growth of an interfacial defect from the inner side may occur (mode II). In fact, defects are already present at the interface or, in the case of perfectly clean interfaces, may be formed as a consequence of a stress aided thermally activated process [44].

By following Carbone et al. [44], we can estimate the critical stress, which makes growing a defect of size $2a$ as

$$\sigma_{II} = \left(\frac{E^* \gamma}{2\pi a} \right)^{1/2} \quad (8.7)$$

when the defect is present at the interface, or as

$$\sigma_{II} = \left(\frac{\pi^3}{12} \frac{\gamma^3 E^*}{U_B} \right)^{1/4} \quad (8.8)$$

when the defect is formed as a consequence of a stress aided thermally activated nucleation process. In Eq. (8.8) U_B is the energy barrier against propagation, depending on the rate of thermal fluctuations through the Maxwell–Boltzmann equation. So if the stress in the walls exceeds the critical stress σ_{II} (or eventually σ_{III}), then the walls structure detaches, even though the conditions for unstable propagation of the crack from the edge do not occur.

Therefore, the detaching moment M_{II} corresponding to mode II debonding mechanism, i.e., the moment for which the normal stress in the wall matches the critical stress σ_{II}, can be easily evaluated taking into account that the system is assumed to be linear elastic, so the stress σ in the wall linearly changes with the applied moment M and, hence,

$$M_{\mathrm{II}} = \frac{\sigma_{\mathrm{II}}}{\sigma} M \tag{8.9}$$

Similarly, the detaching moment M_{III}, which corresponds to mode III debonding mechanism and occurs when the stress exceeds the theoretical interfacial strength $\sigma_{\mathrm{III}} = \gamma/\rho$, ρ being the range of van der Waals forces, can be evaluated by

$$M_{\mathrm{III}} = \frac{\sigma_{\mathrm{III}}}{\sigma} M \tag{8.10}$$

For soft adhesives, the stress σ_{II} is always much less than the theoretical contact strength σ_{III}, $\rho \approx 1$ nm being the typical range of van der Waals forces. Hence, the mode III debonding mechanism cannot occur in real cases. In fact, for $\rho \approx 1$ nm and $\gamma = 16$ mJ m^{-2} we have $\sigma_{\mathrm{III}} = 16$ MPa.

Taking into account that, even for very clean surfaces, the presence of small interfacial defects cannot be excluded, if we assume a defect of size $2a = 20$ nm, the corresponding debonding stress σ_{II} can be calculated by Eq. (8.8), giving $\sigma_{\mathrm{II}} \approx 1$ MPa, which is significantly smaller than σ_{III}. As a result, the critical moment for detachment will be $M_c = \min\{M_{\mathrm{I}}, M_{\mathrm{II}}\}$ and the optimal geometry of the system will be the one maximizing M_C.

Figure 8.5 shows the critical bending moment M_C necessary to fully detach the wall-like structure as a function of the geometrical parameters: (i) film thickness t (Fig. 8.5a), (ii) wall spacing λ (Fig. 8.5b), (iii) wall height h (Fig. 8.5c) and wall thickness s (Fig. 8.5d). Results are normalized with respect to the debonding moment M_{flat} necessary to detach the flat control sample.

The debonding moment M_{I} is maximized when the compliance of the system is increased. However, this not always corresponds to an increase of the critical moment, being in some cases M_{II} less than M_{I}. Therefore, we usually can find an optimal value of the geometrical parameters so to maximize M_C.

For example, it is not convenient to increase λ above a certain value (which, for the example case shown in Fig. 8.5 is about 62), because even though the debonding moment M_{I} increases, the stress in the walls also increases and the mode II debonding mechanism becomes more critical ($M_{II} < M_I$).

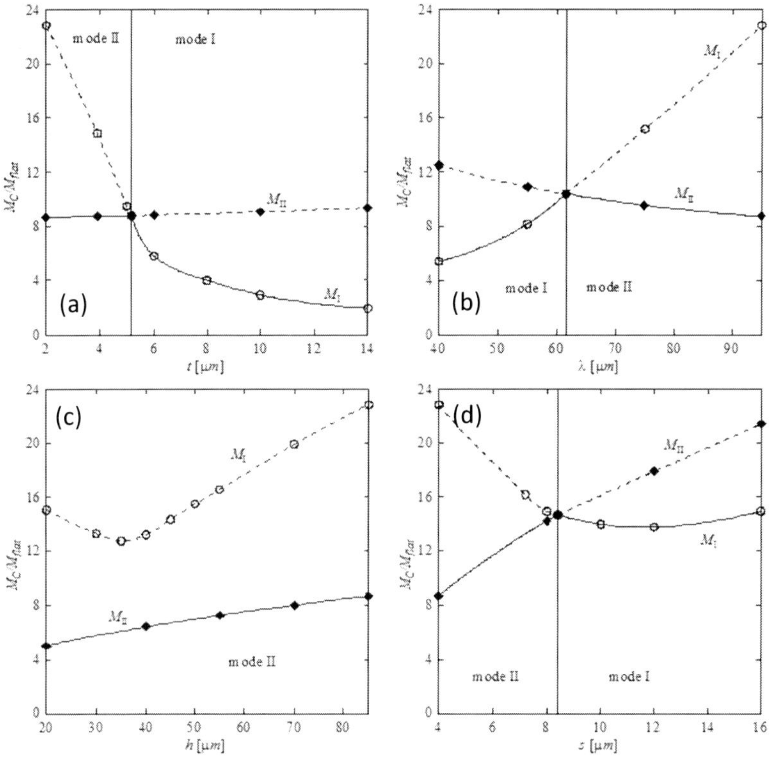

Figure 8.5 Variation of the critical bending moment M_C, necessary to fully detach the wall-like structure, normalized with respect to that of a flat control sample M_C/M_{flat}, with the geometrical parameters: (a) film thickness t (s = 4 μm; h = 85 μm; λ = 95 μm); (b) wall spacing λ (t = 2 μm; s = 4 μm; h = 85 μm); (c) wall height h (t = 2 μm; s = 4 μm; λ = 95 μm); (d) wall thickness s (t = 2 μm; h = 85 μm; λ = 95 μm). Material constants: E (Young's modulus) = 3 MPa; ν (Poisson's ratio) = 0.5. Work of adhesion: γ = 16 mJ/m^2; defect size (considered for evaluating M_{II}) $2a$ = 20 nm.

Moreover the risk of buckling for the walls would need to be considered when defining the optimal parameters, because such risk grows when the microwalls thickness reduces and their spacing and height increase.

8.3.2 Detachment Occurring Obliquely to the Microwalls

For crack propagating obliquely to the wall like structures, the crack trapping phenomenon cannot occur because, as mentioned above, the energy release rate G remains constant during the crack propagation.

In this case, the system cannot more be simplified into a 2D structure. However, the critical moment for detachment can be evaluated by performing an analysis on a single row, and applying convenient boundary conditions on the symmetry planes (see Ref. [51]). In this case 3D solid elements need to be used to carry out the FE analysis.

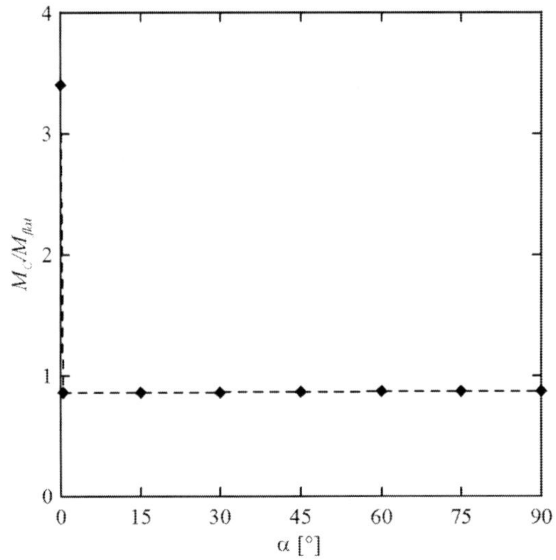

Figure 8.6 The critical bending moment M_C, necessary to fully detach the microstructured surface, normalized with respect to that of the flat control sample M_{flat}, as a function of the angle α between the wall-like structures and the crack front. Work of adhesion: γ = 16 mJ/m^2 (h = 100 micron, λ = 100 micron, s = 10 micron, t = 5 micron).

Moreover, it is possible to demonstrate (see Ref. [51]) that the crack front of propagation is always perpendicular to the bending plane. In fact, only in this case, the crack propagation is stationary and the angle that the crack makes with the direction of the applied moment does not change during the propagation.

In this case, variations of the energy release rate G with the angle α between the crack front and the wall-like structures are negligible. As a result, the critical moment for detachment will be an almost constant function of α. This behavior is confirmed in Fig. 8.6, where the critical moment, normalized with respect to the value obtained for flat surfaces, is plotted as a function of α. Notice the critical moment is larger than the one of the flat control sample only at $\alpha = 0$, i.e. when detachment occurs perpendicularly to the walls direction. In all the other cases, the system is even easier to detach than the flat control sample.

8.4 Conclusions

A microstructured surface like that reviewed in this chapter would be useful in all those engineering applications, (e.g., robot-locomotion and object manipulation) where the adhesion force acts in a preferential directions and no adhesion is required along the other directions to facilitate system detachment. In fact, such a system would have the advantage of guaranteeing high detaching strength when loaded with a moment acting perpendicularly to the wall-like structures, and to be more easily detached from the substrate when loaded with a moment acting in any other direction.

Such considerations are well supported by results of numerical calculations performed on the system.

Acknowledgment

Useful discussions with Prof. G. Carbone are acknowledged.

References

1. Martines, E., Seunarine, K., Morgan, H., Gadegaard, N., Wilkinson, C. D. W., and Riehle, M. O. (2005). Superhydrophobicity and superhydrophilicity of regular nanopatterns, *Nano Lett.*, **5**, 2097–2103.
2. Afferrante, L., and Carbone, G. (2010). Microstructured superhydrorepellent surfaces: effect of drop pressure on fakir-state stability and apparent contact angles, *J. Phys.: Condens. Matter*, **22**, 325107.

3. Geim, A. K., Dubonos, S. V., Grigorieva, I. V., Novoselov, K. S., Zhukov, A. A., and Shapoval, S. Yu. (2003). Microfabricated adhesive mimicking gecko foot-hair, *Nat. Mater.*, **2**, 461–463.
4. del Campo, A., Greiner, C., and Artz, E. (2007). Contact shape controls adhesion of bioinspired fibrillar surfaces, *Langmuir*, **23**, 10235–10243.
5. Autumn, K., Liang, Y. A., Hsieh, S. T., Zesch, W., Chan, W. P., Kenny, T. W., Fearing, R., and Full, R. J. (2000). Adhesive force of a single gecko foot-hair, *Nature*, **405**, 681–685.
6. Autumn, K. (2006). Properties, principles, and parameters of the gecko adhesive system. In *Biological Adhesives* (ed. Smith, A., and Callow, J.), pp. 225–256. Berlin, Germany: Springer.
7. Autumn, K., Dittmore, A., Santos, D., Spenko, M., and Cutkosky, M. (2006). Frictional adhesion: a new angle on gecko attachment, *J. Exp. Biol.*, **209**, 3569–3579.
8. Autumn, K., Hsieh, S. T., Dudek, D. M., Chen, J., Chitaphan, C., and Full, R. J. (2006). Dynamics of geckos running vertically, *J. Exp. Biol.*, **209**, 260–272.
9. Autumn, K., Majidi, C., Groff, R., and Dittmore, A. (2006). Effective elastic modulus of isolated gecko setal arrays, *J. Exp. Biol.*, **209**, 3558–3568.
10. Autumn, K., and Hansen, W. (2006). Ultrahydrophobicity indicates a nonadhesive default state in gecko setae, *J. Comp. Physiol. A-Neuroethol. Sens. Neural Behav. Physiol.*, **192**(11), 1205–1212.
11. Persson, B. N. J. (2003). On the mechanism of adhesion in biological systems, *J. Chem. Phys.*, **118**, 7614–7621.
12. Arzt, E., Gorb, S., and Spolenak, R. (2003). From micro to nano contacts in biological attachment devices. *Proc. Natl. Acad. Sci. U. S. A.*, **100**(19), 10603–10606.
13. Northen, M., and Turner, K. (2005). A batch fabricated biomimetic dry adhesive, *Nanotechnology*, **16**, 1159–1166.
14. Northen, M. T., Greiner, C., Arzt, E., and Turner, K. L. (2008). A gecko-inspired reversible adhesive, *Adv. Mater.*, **20**, 3905–3909.
15. Bhushan, B. (2007). Adhesion of multi-level hierarchical attachment systems in gecko feet, *J. Adhes. Sci. Technol.*, **21**, 1213–1258.
16. Aksak, B., Murphy, M., and Sitti, M. (2008). Gecko inspired micro-fibrillar adhesives for wall climbing robots on micro/nanoscale rough surfaces, *IEEE Int. Conf. Robot. Autom.*, **23**, 3322–3332.

17. Ge, L., Sethi, S., Ci, L., Ajayan, P., and Dhinojwala, A. (2007). Carbon nanotube-based synthetic gecko tapes, *Proc. Natl. Acad. Sci. U. S. A.*, **104**, 10792–10795.

18. Zhao, Y., Tong, T., Delzeit, L., Kashani, A., Meyyappan, M., and Majumdar, A. (2006). Interfacial energy and strength of multiwalled-carbon-nanotube-based dry adhesive, *J. Vac. Sci. Technol. B Microelectron. Nanometer Struct.*, **24**, 331–335.

19. Tsai, Y., Shih, W., Wang, Y., Huang, L., and Shih, P. (2006). E-beam photoresist and carbon nanotubes as biomimetic dry adhesives. In *Proc. 19th IEEE Int. Conf. on Micro Electro Mechanical Systems*, Istanbul, Turkey, pp. 926–929.

20. Santos, D., Spenko, M., Parness, A., Kim, S., and Cutkosky, M. (2007). Directional adhesion for climbing: Theoretical and practical considerations, *J. Adhes. Sci. Technol.*, **21**, 1317–1341.

21. Kim, S., Spenko, M., Trujillo, S., Heyneman, B., Santos, D., and Cutkosky, M. R. (2008). Smooth vertical surface climbing with directional adhesion, *IEEE Trans. Robot.*, **24**, 65–74.

22. Kamperman, M., and Synytska, A. (2012). Switchable adhesion by chemical functionality and topography, *J. Mater. Chem.*, **22**, 19390–19401.

23. Reddy, S., Arzt, E., and Del Campo, A. (2007). Bioinspired surfaces with switchable adhesion, *Adv. Mater.*, **19**, 3833–3837.

24. Xie, T., and Xiao, X. (2008). Self-peeling reversible dry adhesive system. *Chem. Mater.*, **20**, 2866–2868.

25. Northen, M. T., Greiner, C. Arzt, E., and Turner, K. L. (2008). A gecko-inspired reversible adhesive. *Adv. Mater.*, **20**, 3905–3909.

26. Jeong, H. E., Kwak, M. K., and Suh, K. Y. (2010). Stretchable, adhesion-tunable dry adhesive by surface wrinkling, *Langmuir*, **26**, 2223–2226.

27. Nadermann, N., Ning, J., Jagota, A., and Hui, C.-Y. (2010) Active switching of adhesion in a film-terminated fibrillar structure, *Langmuir*, **26**, 15464–15471.

28. Paretkar, D., Kamperman, M., Schneider, A. S., Martina, D., Creton, C., and Arzt, E. (2011). Bioinspired pressure actuated adhesive system. *Mater. Sci. Eng. C*, **31**, 1152–1159.

29. Murphy, M. P., Aksak, B., and Sitti, M. (2009). Gecko inspired directional and controllable adhesion, *Small*, **5**(2), 170–175.

30. Aksak, B., Murphy, M. P., and Sitti, M. (2007). Adhesion of biologically inspired vertical and angled polymer microfiber arrays, *Langmuir*, **23**(6), 3322–3332.

31. Jeong, H. E., Lee, J.-K., Kim, H. N., Moon, S. H., and Suh, K. Y. (2009). A non-transferring dry-adhesive with hierarchical polymer nanohairs, *Proc. Natl. Acad. Sci. U. S. A.*, **106**, 5639–5644.

32. Yang, S. Y., Carlson, A., Cheng, H., Yu, Q., Ahmed, N., Wu, J., Kim, S., Sitti, M., Ferreira, P. M., Huang, Y., and Rogers, J. A. (2012). Elastomer surfaces with directionally dependent adhesion strength and their use in transfer printing with continuous roll-to-roll applications, *Adv. Mater.*, **24**(16), 2117–2122.

33. Glassmaker, N. J., Jagota, A., Hui, C.-Y., and Kim, J. (2004). Design of biomimetic fibrillar interfaces: 1. Making contact, *J. R. Soc. Interface*, **1**, 23–33.

34. Hui, C.-Y., Glassmaker, N. J., Tang, T., and Jagota, A. (2004). Design of biomimetic fibrillar interfaces: 2. Mechanics of enhanced adhesion, *J. R. Soc. Interface*, **1**, 35–48.

35. Jagota, A., and Bennison, S. J. (2002). Mechanics of adhesion through a fibrillar microstructure, *Integr. Comp. Biol.*, **42**, 1140–1145.

36. Persson, B. N. J. (2003). On the mechanism of adhesion in biological systems, *J. Chem. Phys.*, **118**, 7614–7621.

37. Glassmaker, N. J., Jagota, A., Hui, C.-Y., Noderer, W. L., and Chaudhury, M. K. (2007). Biologically inspired crack trapping for enhanced adhesion, *Proc. Natl Acad. Sci. U. S. A.*, **104**, 10786–10791.

38. Gorb, S., Jiao, Y., and Scherge, M. (2000). Ultrastructural architecture and mechanical properties of attachment pads in Tettigonia viridissima (Orthoptera Tettigoniidae), *J. Comp. Physiol. A*, **186**(9), 821–831.

39. Jiao, Y., Gorb, S., and Scherge, M. (2000). Adhesion measured on the attachment pads of Tettigonia viridissima (Orthoptera, Insecta), *J. Exp. Biol.*, **203**(12), 1887–1895.

40. Gorb, S., and Scherge, M. (2000). Biological microtribology: Anisotropy in frictional forces of orthopteran attachment pads reflects the ultrastructure of a highly deformable material, *Proc. R. Soc. London Ser. B*, **267**, 1239–1244.

41. Daltorio, K. A., Gorb, S., Peressadko, A., Horchler, A. D., Ritzmann, R. E., and Quinn, R. D. (2005). A robot that climbs walls using micro-structured polymer feet, *Proc. of 8th Int. Conf. Climbing and Walking Robots*, pp. 131–138. Berlin, Germany: Springer.

42. Gorb, S. N., and Varenberg, M. (2007). Mushroom-shaped geometry of contact elements in biological adhesive systems, *J. Adhes. Sci. Technol.*, **21**, 1175–1183.

43. Varenberg, M., and Gorb, S. (2007). Shearing of fibrillar adhesive microstructure: friction and shear-related changes in pull-off force, *J. R. Soc. Interface*, **4**, 721–725.

44. Carbone, G., Pierro, E., and Gorb, S. (2011). Origin of the superior adhesive performance of mushroom-shaped microstructured surfaces, *Soft Matter*, **7**, 5545–5552.

45. Carbone, G., and Pierro, E. (2012). Sticky bio-inspired micropillars: Finding the best shape, *Small*, **8**(9), 1449–1454.

46. Varenberg, M., and Gorb, S. (2008). Close-up of mushroom-shaped fibrillar adhesive microstructure: contact element behaviour, *J. R. Soc. Interface*, **5**, 785–789.

47. Qu, L., Dai, L., Stone, M., Xia, Z., and Wang, Z. (2008). Carbon nanotube arrays with strong shear binding-on and easy normal lifting-off, *Science*, **322**, 238–242.

48. Parness, A., Soto, D., Esparza, N., Gravish, N., Wilkinson, M., Autumn, K., and Cutkosky, M. (2009). A microfabricated wedge-shaped adhesive array displaying gecko-like dynamic adhesion, directionality and long lifetime, *J. R. Soc. Interface*, **6**(41), 1223–1232.

49. Afferrante, L., and Carbone, C. (2012). Biomimetic surfaces with controlled direction-dependent adhesion, *J. R. Soc. Interface*, **9**, 3359–3365.

50. Maugis, D. (1999). Contact, Adhesion and Rupture of Elastic Solids, Springer Series in Solid State Sciences, Springer-Verlag, Berlin, Heidlberg, New-York.

51. Afferrante, L., Carbone, G. Grimaldi, G., and Demelio, G. (2014). Micro-textured surfaces with parallel wall-like structures: "Modulation" of adhesion properties with the direction of the applied external moment, to appear in *Proceedings of the ASME 2014 International Mechanical Engineering Congress & Exposition*, November 14–20, 2014, Montreal, Quebec, Canada.

Section IV

Medical Applications

Chapter 9

Bioadhesives as Surgical Sealants: A Review

Leslie John Ray Foster

Bio/Polymer Research Group, Biotechnology & Biomolecular Sciences, University of New South Wales, Sydney, NSW 2052, Australia

J.Foster@unsw.edu.au

The disadvantages of sutures and other mechanical fasteners have led to the development of sutureless technologies. A number of surgical adhesive sealants have been commercialised. The majority of these are advised as adjuncts and still require mechanical fasteners. This chapter provides a review of bioadhesives that have been developed as commercial surgical sealants. Ongoing research using these bioadhesives is presented and used to forecast future developments in the field.

9.1 Introduction

Surgical reconnection of damaged tissues is essential for the restoration of their structure and function. Mechanical fasteners, including sutures, staples and wires remain the most widely used method for joining tissues. However, the first recorded use of a

Bioadhesion and Biomimetics: From Nature to Applications
Edited by Havazelet Bianco-Peled and Maya Davidovich-Pinhas

ISBN 978-981-4463-98-0 (Hardcover), 978-981-4463-99-7 (eBook)
www.panstanford.com

mechanical fastener, namely sutures, is over 4000 years old and despite advances in material and fastener design a number of disadvantages remain [14].

- Mechanical fasteners causes physical trauma to the surrounding tissues and interferes with the functional rehabilitation of the wound site [112]. For example, the closure of traumatic injuries can result in the agglomeration of many small lesions [130], while their use in nerve repair leads to scarring and a lack of conduction [101].
- There is a wide variation in wound healing and eventual tissue scarring due to the different surgical expertise.
- Suturing is an acquired technical skill with significant variation in ability between surgeons. Furthermore, the accuracy of positioning of the mechanical union may be difficult in regions of the body not easily accessible [71].
- Sutures act as a nidus for infection.
- Suturing is a time-consuming procedure.
- A subsequent procedure is usually required at a later date to remove non-degradable mechanical fasteners, increasing the number of follow up visits for the patient and stress [123]. Suture removal increases the risk of infection and wound dehiscence.
- Sutures are not suitable for inherently complicated procedures, such as those with tissues of low cohesion energy, including lung, liver, spleen and kidney.
- Sutures and other mechanical fasteners bring wound edges together rather than sealing wounds; consequently leakage of body fluids or gases may occur.

In comparison to mechanical fasteners, the application of surgical adhesives appears extremely attractive, with faster application, less pain, less traumatic closure and improved tissue regeneration and function. Furthermore, recent advances have reported their potential for localised drug delivery [2,48].

Surgical adhesives are glues capable of adhering tissue to tissue or non-tissue surfaces such as prosthesis. They may also be used to control bleeding (haemostats) or to serve as a barrier for the prevention of body fluid and gas leakage (sealants) [139]. Spotnitz and Burks have recently proposed a classification system for these three surgical tools to aid selection; however, there is a degree of multi-functionality of some components; for example, fibrin-based adhesives are readily used as both sealant and haemostat and has separate FDA approvals for each [140].

In keeping with the focus of this book on bioadhesion, this chapter reviews the principles and application of commercial bioadhesives, as well as the development and potential of new bioadhesives, for surgical sealants.

9.2 Commercial Sealants

There are a number of commercial adhesives approved by various regulatory authorities for medical use and are regularly applied as sealants. Sealants can be divided into those of biological origin, those that are chemical based and a combination of the two. This chapter will focus on bioadhesives as sealants; detailed reviews of synthetic adhesives can be found in references [139,140]. Bioadhesives are considered more attractive as it is generally recognised that sealants based on biopolymers and cross-linked through biochemical reactions can provide more biocompatible alternatives to synthetic, chemical glues [11]. Consequently, bioadhesives, particularly those based on fibrin and collagen are amongst the most commonly used sealants.

Table 9.1 Common commercial bioadhesives used as sealants

Product	Origin/compound	Manufacturer
Fibrin-based		
Tisseel VH®	Human	Baxter
Vivostat®	"	Pacific medical systems
Cryoseal®	"	Thermogenesis
Evicel®	"	OMRIX pharmaceutical
Vitagel®	"	Baxter
TachoComb® & TachoSil®	Human, bovine and equine	Takeda pharmaceutical
Collagen-based		
CoStasis®	Bovine and human	Cohesion technologies

(*Continued*)

Table 9.1 (*Continued*)

Product	Origin/compound	Manufacturer
Gelatine-based		
FloSeal Matrix®	Bovine and human	Baxter healthcare
Gelfoam® Plus & Surgiflo®	Porcine and human	Johnson & Johnson
Albumin-based		
BioGlue™	Bovine + glutaraldehyde	CryoLife
Chitosan-based		
HemCon® Patch PRO & Others	Crustacea	HemCon Medical Technologies Inc.
Syvek®	"	Marine Polymer Technologies Inc.
Celox™	"	Celox Medical
ExcelArrest®	"	Hemostasis LLC Co.
Cellulose-based		
Synaero™	Oxidised cellulose	HemCon Medical Technologies Inc.
Surgicel® & Nu-Knit®	"	Johnson & Johnson

9.3 Fibrin-Based Sealants

9.3.1 Principles

Commercial fibrin sealants emulate the final stages of the coagulation cascade initated upon wound injury and consist of two separately contained components, the "sealant" and the "catalyst". Thrombin (the catalyst) is a serum protease which, in the presence of calcium ions, cleaves the fibrinogen chains (the sealant) into fibrin monomers. Thrombin also activates Factor XIII, also present in the sealant, which promotes the polymerisation of the fibrin monomers into the fibrin clot, i.e. the "seal".

The fibrin clot produced by the sealant degrades by physiologic fibrinolysis in approximately two weeks. The addition of aprotinin, which inhibts the action of human trypsin and plasmin, to the sealant can be used to reduce fibrinolysis therby increasing the stability of the clot. Variations in the concentrations of fibrinogen and thrombin influence the material strength of the clot, while the speed and degree of the reaction is also mainly determined by the thrombin concentration [75,131].

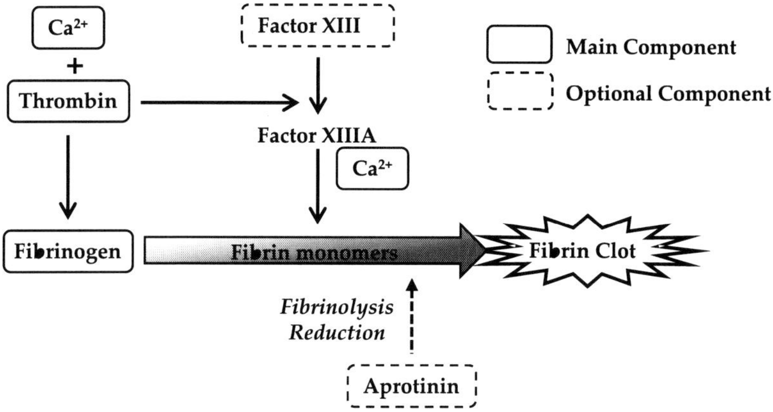

Figure 9.1 Schematic of the fibrin sealing process, with main and optional fibrin components identified. Adapted from Spotnitz (2001) [138].

9.3.2 Commercial Fibrin Sealants

Given that fibrin sealents derive their action from the blood coagulation cascade, it is unsurprising that their main application is as haemostatic sealants [69,137]. However, they have also been used to prevent leakage of cerebro-spinal fluid (CSF) in dura mater repair [130,153] and air leaks in lung procedures [128] and to treat lymphatic leaks in radial neck dissections [56].

The majority of commercial fibrin-based sealants, such as Tisseel VH®, Evicel® and Vivostat®, have formulations derived from pooled human sera [84,121]. However, their derivation from natural blood products poses risks of viral and prion transfer [139,149]. To avoid the problem of possible contamination, virus-inactivated fibrin sealants have now been developed; however, they have a rather complex preparation procedure and this consequently increases the cost of downstream processing for these products [11]. Similarly, safety concerns with some non-human components found in Vitagel®, bovine thrombin, include antibody formation, while Tisseel VH® contains synthetic aprotinin, which may be a potential cause of severe allergenic reactions [149]. Busuttil and Oswald et al. have reported that fibrin sealants with aprotinin can be a cause of anaphylactic shock to the point of death [15,108].

In addition to safety concerns regarding the biological components of fibrin-based sealants, the technology itself has a number of other drawbacks. Fibrin sealants have relatively poor tissue adhesion, about 8 kPa, and a leakage pressure of approximately 13 kPa [81]. Consequently, fibrin sealants are usually used as adjuncts to mechanical fasterners, where they can significantly improve the wound sealing capacities but increase procedural duration [99,131].

Necessary separation of the catalyst and sealant components until point of application means that flowable fibrin sealants are found in combined two syringe systems delivering a liquid or gel-based adhesive to the wound site [139,140]. The gel-like morphology of the sealant combined with a wet operating field ensures that there are variations in delivery and sealing capabilities within the wound site. Thus, fibrin sealants are less effective when applied to wet operating fields [149]. Furthermore, fibrin sealants such as Tisseel VH® require pre-warming to body temperature as well as a significant curing time, which further increases the duration of procedures and chances of infection [131].

TachoComb™ and TachoSil™ attempt to overcome the hurdles of non-uniform fibrin sealant delivery and tissue bonding observed with other commercial sealants, through a film delivery system. These film sealants consist of a thin layer of lyophilised human fibrinogen and bovine thrombin on one side of an equine collagen sheet [38]. While these systems have been successfully used in haemorrhage control in parenchymateus organs, there are currently insufficient clinical comparisons to other fibrin sealants.

9.3.3 Research and Development

Continued development of fibrin sealants have focused on systems for more easy application, as well as better and more uniform sealing in wet operating fields. A number of authors have reported the development of fibrin-based bandages. Anema et al. used concentrated mixtures of lyophilised fibrinogen and thrombin on a polyglactin backing, this system was easy to apply and supported rapid haemostasis but formed a gelatinous clot when applied [6]. Larson et al. [78], Cornum et al. [25], and Holcomb et al. [66] have all reported dry fibrin bandages that significantly reduce both short and long-term blood loss during haemorrhage. However, despite their efficiency, dry fibrin sealants remain expensive and are relatively stiff prior to application [113]. Similarly, sprayable-foam fibrin sealants have also been formulated with apparently superior performance to their liquid counterparts; however, their potential applications remain restricted as haemostats [132].

Other research has investigated the inclusion of various agents to existing fibrin sealants in attempts to improve haemostatic performance and reduce cost. For example, Rothwell and co-workers report that while the addition of a proprietary propyl gallate-based procoagulant to fibrin bandages supported haemostasis, the inclusion of ε-amino caproic acid actually increased blood loss [125,126]. Commercially, both cellulose (Surgicel®) and gelatine-thrombin (Floseal®)-based sealants containing fibrin have been formulated to reduce cost.

New cross-linking mechanisms may provide an alternative avenue to improve the relatively poor tissue adhesion and operating strength of fibrin sealants [12,32,33]. Initial studies by Elvin et al. using a combination of ruthenium and blue light suggest a product with a more rapid application and curing as well as greater tissue adhesion strength [30,33].

The reliance upon pooled human plasma for the production of fibrin-based sealants presents the risk of viral and pathogen transmission [67]. Sawyer and co-workers have attempted to overcome this issue by using fibrinogen and thrombins derived from salmon blood to produce bandage sealants [100,127,152]. These salmon-derived sealants were highly effective in controlling haemorrhage in a severe swine aorta bleeding model [100]. However, concerns regarding the direct effects of thrombin in patients' blood as well as potential antibody responses remain.

9.4 Collagen-Based Sealants

9.4.1 Principles

Collagen sealants are usually augmented with thrombin and require fibrinogen in the patients own plasma and therefore, similar to fibrin-based sealants, also takes advantage of the body's coagulation cascade. Commercial collagen-based surgical sealants physically trap blood and its coagulation products in dense fibrous mats, while the collagen also induces platelet adhesion and aggregation, and activates blood coagulation factors [41]. Van der Meijden et al. have recently shown that collagen type I provoked a dose-dependent reduction in the clotting time of human plasma through activation of factor XII, which bound to the collagen, stimulated plasma generation of thrombin and increased the effects of platelets through activation of glycoprotein VI [151].

9.4.2 Commercial Collagen Sealants

While collagen-based sealants are relatively new, a number of commercial products have found their way to market. CoStasis/DynaStat® is a sprayable liquid approved by the FDA as a haemostatic sealant and consists of bovine collagen and thrombin, augmented with the patients own plasma, a separate product, the "CellPaker® Plasma Collection Device" is required [26]. "FloSeal" and "Proceed" were originally developed by Fusion Technologies Inc. for applications in vascular haemostasis and sealing CSF leaks, respectively [122]. Similar to CoStasis®, both products were produced using a combination of bovine thrombin and collagen. However, since the takeover of Fusion Technologies Inc. by Baxter, FloSeal Matrix® is now produced with human thrombin and bovine gelatine and advised as a haemostat [46].

A controlled trial with over 300 patients demonstrated the success of CoStasis® in reducing blood loss when used in tumor resections, abdominal procedures, muscle flap harvesting for reconstructive purposes, reoperative exploratory surgery and trauma-related procedures [19]. However, more studies on the safety of CoStasis® for neurological, ophthalmic and urological procedures are required [46,64].

Compared to fibrin sealants, the use of bovine thrombin and collagen reduces the chances of viral infection, although prion transfer may still give cause for concern. Similarly, collagen-based sealants may be comparatively less expensive than their fibrin competitors [121]. However, as with fibrin sealants, collagen-based sealants are usually adjuncts to mechnical fasteners and liquid or gel-based, which limits their appication while the disadvantages associated with the mechnial fasteners remain.

9.4.3 Research and Development

Reliance upon mechanical fasteners due to a comparatively low tissue adhesion strength remains an issue, with researchers attempting to use new cross-linking techniques while retaining biocompatibility. For example, Tanaka and co-workers have used novel citric acid–derived cross-linkers to enhance the bonding strength of collagen sealants [146,147]. The tissue adhesion strength of this new sealant was reported to be 11 times greater than that of a fibrin sealant. Taguchi et al. report that the sealant also showed good biocompatibility and resorbability [147]. However, a cure time of approximately 10 min for this sealant represents a significant hurdle for its clinical application [147].

More recently, Baik et al. report the use of a complex mix of non-immunogenic, esterified atelocollagen with 3,4-dihydroxyphenyl-L-alanine (DOPA) [9]. High proportions of DOPA residues were found in proteins responsible for the attachment of Mussels (*Mytilus edulis*) to marine surfaces and these rediceus have been demonstrated to be critical for this adhesion [111,150]. While this collagen/DOPA sealant showed signifcant adhesion and haemostasis within 3 min, the strength was compromised, being approximately similar to fibrin sealant [9].

9.5 Gelatine-Based Sealants

9.5.1 Principles

Gelatine is a mixture of proteins and peptides produced by the partial hydrolysis of collagen. As such, gelatine-based sealants are a logical extension of their collagen counterparts, with the advantage of being able to form gels in situ, making them ideal for soft tissue applications [89]. Gels are formed through the entanglement of the long proptein and peptide chains and through the formation of cross-links between the chains. The physical gelation process of gelatine, where triple helices create temperature-dependent reversible connections between the chains, is well known [30]. In contrast, commercial sealants utilise body enzymes such as thrombin to catalyse the chemical gelation of gelatine, whereby permanent bonds between specific amino acids on the gelatine chains are created, effectively sealing the gel structure [30].

Gelatine granules swell by about 20% upon contact with blood and other fluids within a period of about 10 min. Tissue adhesion is obtained in the same manner as with collagen sealants [46].

9.5.2 Commercial Gelatine Sealants

Baxter's FloSeal Matrix® consists of cross-linked bovine gelatine granules and human thrombin and is FDA advised as a hemostatic sealant. The product also provides a small bowl and requires the mixing of the thrombin to the gelatine granules prior to application [46]. Mixing of the human thrombin catalyst, "Thrombin-JMI", and porcine gelatine sealant is also required with Johnson & Johnson's Surgiflo® [143]. Both commercial sealants are used as topical hemostats in various anatomical sites and surgical procedues, but are not for application in ophthlamic or urologic surgery [109]. Woodworth et al. have demonstrated the success of Surgiflo® for haemostasis in endoscopic sinus surgery, reporting a 96.7% success rate within 10 min and a median time of just over a minite when accompanied by physical compression [155]. Both sealants are readily bioresorbable within a 4–6 week period [46,143].

Thrombin from pooled human plasma has the potential risk of viral and prion-related disease transmission due to the multiple units of blood used in manufacturing each product lot [53]. In contrast, antibody formation associated with the use of thrombin from bovine sources can lead to coagulopathy and possibly death. Consequently, the FDA has imposed its Black Box Warning, its highest level of concern [148]. Recombinant thrombin reduces the risk of known disease transimission as well as antibody formation. However, allergic reactions to hamster and snake proteins use in the manufcacture of recombinant thrombin are potential dangers [120].

Commercial bovine and recombinant thrombin are provided as lyophilised powders requiring reconstitution [40,148]. In contrast, thrombin from human plasma is supplied mixed but requires thawing prior to immediate use with limited refrigerated storage capabilities [53]. Gelfoam® Plus provides a porcine gelatine sponge combined with lyophilised pooled human thrombin in a single package. The kit has the added advantage that it does not require refrigeration [53].

9.5.3 Research and Development

There has been considerable research on various cross-linking mechanisms for gelatine gels, including self-assembling peptides [57], enzymatic cross-linking and stimuli-responsive materials [72]. For example, microbial transglutaminase (mTG) has been reported to cross-link gelatine; this gelatine-mTG adhesive could bond to wet tissue with a greater adhesive strength than fibrin sealant [21,22]. However, the potential application of these biomedical gels as adhesive surgical sealants has not yet been satisfactorily explored.

A number of researchers have formulated sealants where aldehydes are utilised to chemically cross-link gelatine [91,94,132]. For example, Ikada et al. report that increases in glutaraldehyde (GA) concentration in cross-linking gelatine increased the adhesion strength of GA-aldehydre films to porcine skin [94]. Manju et al. used an alginate dialdehyde cross-linked hydrogel to seal PET fabric intended as vascular grafts, in the process reducing the haemolytic potential and improving cellualr proliferation [91]. More recently, Suzuki and Ikada have emphasised the effectiveness of glutaraldehyde in improving gelatine cross-linking [144]. This natural-synthetic hybrid sealant exhibited a tissue bonding strength almost three times that of fibrin glue and double the leakage pressure; sealant properties were dependent upon component concentrations [144].

Photoactivation offers another route to improve gelatine cross-linking and sealant properties. Nakayama and Matsuda partially derivatised bovine gelatine with UV reactive benzophenone before combining it with poly(ethylene glycol) diacrylate (PEGDA) and ascorbic acid to fabricate an adhesive sealent. After activation through low-powered (5 mW cm^{-2}) irradiation with a pulse laser, this sealant exhibited an in vitro adhesive strength to collagen films of about 12 kPa, compared to 1.5 kPa for commercial fibrin sealant under the same conditions [103]. When tested for liver haemostasis, the sealant showed neglibible tissue damage and eventually resorbed [103]. As the use of UV irradiation to effect a cross-linking process between gel and tissue has the risks of increasing cellular DNA mutation, Matsuda and co-workers also developed a light-activated gelatine-based sealant [88]. Styrene-derivatised gelatine, PEGDA and carboxylated camphorquinone irradiated with blue light was used to seal incised abdominal rat aortas and gradually degraded within 4 weeks [88,106].

Rapid photochemcial cross-linking using ruthenium and blue light has been applied to fish, bovine and porcine gelatine to produce highly flexible tissue sealants with adhesion strengths increasing from 57 to 110 kPa [34]. This method avoids the need for derivitisation by covalently cross-linking its constituent resilin and fibrinogen proteins with tyrosine (d-Tyr) [42]. These dityrosine-cross-linked sealants effectively sealed lung, vascular and gastrointestinial defects with minimal inflammation [70]. Furthermore, when combined with fibrinogen, the sealant actively promoted cell growth in vitro, suggesting its additonal potential for localised drug delivery [129].

Despite extensive research aimed at the formulation of gelatine-based tissue sealants, various studies have highlighted their limitations regarding their speed of polymerisation, elasticity, tissue adhesion strength and cytotoxicity [22,55,68,96,142].

9.6 Albumin-Based Sealants

9.6.1 Principles

CryoLife's BioGlue™ is a combination of purified bovine serum albumin (BSA) and glutaraldehyde, dispensed through a controlled delivery system. At the tissue interface, adhesion is mediated through sealant cross-linking to tissue amines and consequently adhesion strength is regulated by aldehyde content and its affinity to tissue type [8]. BioGlue™ also adheres to synthetic graft biomaterials through mechanical interlocks within the interstices of the graft matrix [10].

9.6.2 Commercial Albumin Sealants

With BioGlue™, stoichiometrically equivalent doses of glutaraldehyde (10%) and albumin (45%) are admixed from a custom cartridge delivery system. It is purported to be the leading surgical adhesive in cardiovascular procedures [10,147]. This sealant has distinct advantages in that it requires no preparation and has a long shelf life without the need for refrigeratation. Avoidance of pooled human plasma origins reduces the risk of viral transmission, but the use of bovine albumin and a synthetic cross-linking agent maintain concerns regarding hyperallegenicity and toxicity.

The tissue adhesion strength for BioGlue™'s advised cardiovascular application is insufficent to sutureless application and consequently this commercial sealant is FDA indicated as an adjunct to mechanical fasteners. Various authors have also reported the application of BioGlue™ in a variety of procedures, including nephron-sparing laparscopy [63], treatment of anal fistulas [28] and dural repair [7]. However, serious adverse effects, including stroke, infection, meningitis and CSF leaks have been reported when applied in neurosurgery [4,10]. LeMaire et al. report that BioGlue™ reinforcement in pediatric aortic surgery impairs vascular growth and caused stricture when applied circumferentially around an aorto–aortic anastomosis [86]. Similarly, Klimo et al. report a variety of wound complications associated with the application of BioGlue™ in neurosurgical procedures in pediatric patients [76]. Consequently, BioGlue™ is currently advised for use in adult patients only [10]. Furthermore, leakage of BioGlue™ through needle holes in aortic tissue and prosthetics have raised concerns about embolisation [87]. In a 2006 study, BioGlue™ was used to repair transphincteric fistulas with initially promising results, but led to prolonged severe pain, significant postulant discharge and abscess formation, to such an extent that the trial was terminated early [3].

9.6.3 Research and Development

Research on albumin-based sealants have focused on the verstaility of BioGlue™ in clincial scenarios. Results from a number of studies highlight increasing concern with the application of such tissue sealents where chemical cross-linking agents are utilised, while also emphasising the need to design adhesive sealants for particular clinical applications [3,4,7,8,10,16,29,63,86,87,117].

While not classified as sealants, it should also be noted that albumin solutions have been readily utilised to agument laser tissue welding (LTW) in sealing wounds. During LTW, the albumin solder is applied to severed tissue ends and irradiated using an infra-red laser. As the albumin "solder" reaches a temperature of approximately 70°C, the proteins are denatured and intertwine with tissue collagen fibres [98]. Albumin solders have been successfully demonstrated to prevent fluid leakage in urethral reconstruction [115] and applied in the repair of hypospadias [74], the aorta and carotid arteries [90].

Despite new feedback and robotic systems to control laser power and tissue temperature, protein denaturation and thermal damage to tissue can still occur during solder-assisted LTW [45,102,77]. In an effort to reduce irradiation energy and avoid tissue injury, Lauto et al. used a non-cytotoxic, natural protein cross-linking agent, genipin [79]. Inclusion of genipin in the albumin solder, doubled the tissue adhesion strength (~17 kPa), the authors speculate that the genipin may have supported the sterical intertwining of the collagen fibres by reducing solder solubility, as well as possible cross-linking of the protein with the collagen through its amino groups [79].

9.7 Chitosan-Based Sealants

9.7.1 Principles

Chitosan, poly(-*N*-acetylglucosamine), is produced commercially through the chemical or enzymatic deactylation of chitin, the main biopolymer in the exoskeletons of crustacea and insects [111], and directly though bioprocessing of fungi such as *Mucor rouxii* and *Aspergillus niger* [17,20,114]. As the effectiveness of the deactylation process can be manipulated but is never complete, chitosans are a random distribution of chitin and chitosan monomeric units, in various ratios with the latter in majority [27]. It is readily used in a variety of commercial applications, including various foodstuffs [1]. Chitosan is also adhesive to various tissues. Amino groups in chitosan bond to tissue collagen through electrostatic attraction, hydrogen bonds and hydrophobic effects; reductions in number of amino groups reduces their ability to adhere to tissue [136].

The haemostatic property of chitosan has been well documented; the biopolymer not only promotes blood coagulation in vivo, but also enhances the release of platelet-derived growth factor (PGDF-AB) and transforming growth factor (TGF-β1), important in the wound healing process [105]. Work by Rao and Sharma suggests that the haemostatic mechanism of chitosan is independent of the coagulation cascade and more likely due to a direct interation between the biomaterial and erythrocyte cell membranes [118].

9.7.2 Commercial Chitosan Sealants

A range of chitosan-based dermal dressings have been commercially produced. HemCon® Patch PRO is a non-invasive haemostatic patch that enables medical professionals to enhance their direct pressure protocol during catheterisation procedures, reducing time to haemostasis [24,44]. HemCon Medical Technologies Inc. (a subsidiary of TriStar Wellness Solutions Inc. as of May 2013), have also produced a range of commercial chitosan haemostat products, including bandages, X-ray detectable temporary surgical dressings, gauze and dental dressings [61]. Similar chitosan-based haemostatic fibrous sealants have also been produced by Marine Polymer Technologies Inc. (Syvek®) and Celox Medical (Celox™ Gauze Z-fold) [18,92]. Furthermore, a range of companies have released soluble, particulate chitosan as haemostats, including ExcelArrest® (Hemostasis LLC Co.) and Celox-A™ [18,41]. Commercial chitosan-based sealants have been used as haemostats for the treatment of severe trauma [13,95,154], and parenchymal and large venous haemorrhage in a swine model [116].

Dependent upon the degree of acetylation, molecular weight and morphology, chitosan exhibits varying antimicrobial properties [49,119], and HemCon Medical Technologies Inc. also claim that the bandages provide a strong antibacterial barrier even against antibiotic-resistant organisms, including MRSA (methicillin-resistant *Staphylococcus aureus*) and VSA (vancomycin-resistant *Enterococcus faecalis*) [62].

Chitosan is officially derived from animal tissues and consequently commercial products still require official controls for sourcing and the "elimination and/or inactivation of viruses and transmissible agents" [35–37]. Commercial production of "ultra-pure, animal free GMP chitosan", such as that by KitoZyme®, may help reduce costs associated with current standard compliance[73].

9.7.3 Research and Development

The development of chitosan-based haemostats continues to be explored. For example, Dowling et al. have recently reported the production of a amphiphillic, chitosan-based sealant which self-assembles to gel blood, but also supports subsequent reversible action with the release of cyclodextrins [31].

Chitosan derivatised with azide and lactose formed an insolubale hydrogel after a minute's irradiation with UV light. The tissue binding strength of this hydrogel was similar to fibrin glue but exhibited better sealing qualities, with respective air leakage pressures of 8.1 and 6.8 kPa from intestine, 26.7 and 10.7 kPa for the aorta, and 10.1 and 5.6 kPa from the trachea [93,107]. More recently, Hattori et al. have used UV irradition to produce a photo-cross-linkable hydrogel of chitosan and calcium alginate for application as a haemostat [60].

In an effort to overcome the disadvantages associated with commercial gel-based sealants, Foster and co-workers impregnated insoluable chitosan films with an FDA approved chromophore, indocynanine green (ICG) [50,80]. This flexible thin film could be readily cut to size and easily applied to anastomose severed nerve ends [82,83]. The ICG adsorbed near infrared (nIR) irradiation to produce a tissue binding strength signficantly greater than that of fibrin glue (~18 kPa) while avoiding the need for mechanical fasteners [50,81]. Furthermore, the use of nIR overcomes the potential mutagenic effects of UV, while the low power used (120 mW) caused no tissue damage [82,83]. This biodegradable, chitosan-based adhesive signficiantly reduced operating times, and has been successfully used to seal intestinial and corneal wounds with leakage pressures exceeding 80 kPa [80,131]. The ICG does not act as a cross-linking agent, but appears to focus the laser energy to support the expansion of chitosan chains, therby increasing the availability of amide groups bonding to collagen fibres [131]. More recently, Foster et al. demonstrated that this thin-film adhesive could also be used to locally deliver antibiotics, also avoiding their refrigerated storage, with an emphasis in preventing and treating ocular infections [48]. Similarly, Garcia et al. have reported the use of chitosan films in robotic LTW to seal scleral wounds [51].

9.8 Cellulose-Based Sealants

9.8.1 Principles

The natural affinity of cellulose, the most abundant biomass on earth, for self-adhesion has been long recognised and various theories have been postulated based on mechanical interlocking, diffusion, adsorption and possible weak boundry layers [52].

Oxidation of cellulose introduces a high carboxyl content, and supports its modification into an adsorbable biomaterial; control of the oxidation reaction permits manipulation of the physical properties and degradation characteristics [141]. Once implanted this fibrous biomaterial imbibes surrounding fluid and loses much of its material strength, becoming gelatinous within 8 hours and 80% being adsorbed within a few days [141]. The haemostatic action of oxidised cellulose is mainly due to mechanical action as the fibrous matrix traps blood cells and supports the formation of a sealing clot [141].

9.8.2 Commercial Cellulose Sealants

HemCon Medical Technologies Inc. have recently released Synaero™ Hemostatic Gel indicated for haemostasis nasal surgery. Its application onto the nasal mucosa through a syringe and provides haemostasis without the need for packing, thus providing clear visibility and accessibility to the operating field, supporting faster procedures and reducing patient discomfort [145].

Davidson et al. reported that Surgicel®, a haemostat gauze produced from oxidised cellulose, was as effective as Vivostat®, a patient-derived fibrin sealant, in achieving haemostasis during partial hepatectomies [28]. In contrast, Hanks et al. report that a randomised, multicenter clinical study showed that the patient-derived fibrin sealant, Vivostat®, performed significantly better in a variety of procedures than the cellulose-based sealant Surgicel®; Vivostat® required 1.6 min to achieve haemostasis, while Surgicel® needed 3.3 min [58].

9.8.3 Research and Development

Research on the clinical use of cellulose sponges sealed with fibrin-based adhesives has led to the inclusion of cellulose as a biomaterial within surgical sealants [5]. For example, Finley et al. report the clinical combination of oxidised cellulose (Surgicel®) with fibrin glue (Tisseel®) in a "sandwich" for laparoscopic wedge resection of small renal lesions, showing excellent haemostatic action without the need for clamping [45]. Similarly, Germanos et al. used a combination of Surgicel® with FloSeal® to prevent venous bleeding during rectal surgery [54]. More recently, the use of cellulose has been explored as a barrier to the formation of adhesions [23,124].

Haznedaroglu and co-workers have recently reported the development of a new haemostat, "Ankaferd Blood Stopper" (ABS), based on a folkloric medicinal plant extract [39]. This ABS demonstrated good haemostatic action in both cardiac and renal surgery [60,110].

9.9 Summary and Future Trends

Mechanical fasteners in surgery represent an ancient technology. Sutures remain the gold standard in wound closure by virtue of their strength, but their disadvantages have driven research and development towards sutureless technologies. With the global market for wound closure products expected to reach US$ 4.6 billion by 2016, the growth in high strength adhesives and energy-based modalities, such as LTW, are expected to double, claiming a greater market share [97].

Amongst other disadvantages, mechanical fasteners fail to seal wounds; commercial surgical sealants play a vital role in haemostasis and permit surgeons greater visibility in the operating field. Currently, the majority of commercial sealants utilise the body's natural coagulation cascade to promote the formation of a sealing blood clot. While fibrin-based sealants (Tisseel®) are the most commonly used, other companies have substituted fibrin for less costly and safer biopolymers with some adhesive properties, such as collagen (CoStasis®) and gelatine (FloSeal Matrix®). Nevertheless, a reliance upon thrombin, as a clotting catalyst, has raised issues from viral and prion transfer to severe allergic reactions. The tissue adhesion strengths of these technologies are relatively limited and consequently they are FDA advised as adjunct haemostats, to be used in combination with mechanical fasteners. Thus, in pursuit of sutureless technology, industry has introduced chemical cross-linking agents in combination with these biopolymers (e.g. BioGlue) to increase tissue adhesion. However, the use of synthetic agents has raised concerns regarding their toxicity and allergic reactions, limiting their application. Following this trend, the strongest commercial surgical adhesives with sealing capabilities are presently synthetic with no "bio" component, such as the cyanoacrylates (e.g. Histoacryl®, Braun

Scientific) and PEG-based hydrogels (e.g. CoSeal™, Cohesion Technologies). Thus, while sealants are routinely used in clinical procedures, active questions and limitations force physicians to choose between extremes of adhesion strength and biocompatibility.

Researchers have continued to explore the clinical versatility of current haemostats as well as the development of new high strength bioadhesives to seal wounds and free surgery from the shackles of mechanical fasteners. Chemical and photo-activated cross-linking of synthetically derivatised biopolymers have shown considerable promise for soft tissue applications.

The majority of sealants being investigated are fluid or gel systems, which are difficult to apply in wet operating fields and have inherent problems in delivering a uniform bonding to tissue. Industry has responded with the development of haemostatic bandages or films as adjunct to mechanical fasteners (HemCon® Patch Pro). Laser- or light-activated bioadhesives (LAB) in a flexible, thin-film morphology exhibit more uniform sealing and support minimally invasive surgery, while expanding potential applications beyond haemostasis to nerve, intestine and ocular repair.

While synthetic and natural-synthetic hybrid surgical sealants continue to be explored, ongoing research is investigating more natural solutions using biomimetic strategies. Bioadhesion mechanisms exhibited by marine organisms such as mussels, barnacles, tube worms, starfish and algae are providing new adhesive compounds such as DOPA [11]. Similarly, the ability of gecko feet to adhere to smooth and rough, vertical and inverted surfaces has promoted the development of a gecko-mimetic adhesive [84].

Surgical bioadhesives and sealants have been developed with function and biocompatibility as key characteristics. Increasingly, researchers are exploring means by which these bioadhesives can support functional tissue regeneration. Bioadhesive sealants also offer means of local drug delivery, thus reducing cost and whole body effects. Continued research on the ability of surgical bioadhesive sealants to deliver pharmaceutical agents to support biocompatibility, promote tissue regeneration and prevent infection is ongoing.

Acknowledgements

The author acknowledges financial support from the National Health and Medical Research Council of Australia (NHMRC, APP630510) and the Ophthalmic Research Institute of Australia (ORIA, 2009).

References

1. Agullo, E., Rodriguez, M. S., Ramos, V., and Albertengo, L. (2003). Present and future role of chitin and chitosan in food, *Macromol. Biosci.*, **3**(10), 521–530.
2. Ahmed, T., Marcal, H., Johnson, S., Brown, B. N., and Foster, L. J. R. (2012). Coalescence of extracellular matrix (ECM) from porcine urinary bladder (UBM) with a laser-activated chitosan based surgical adhesive, *J. Biomater. Sci: Polym. Ed.*, **23**(12), 1521–1538.
3. Alexander, S. M., Mitalas, L. E., Gosselink, M. P., Oom, D. M. J., Zimmerman, D. D. E., and Schouten, W. R. (2008). Obliteration of the fistulous tract with BioGlue adversely affects the outcome of transanal advancement flap repair, *Techniq. Coloproct.*, **12**(3), 225–228.
4. Aloke, S., Green, K. M. J., Khan, M. I. J., Saeed, S. R., and Ramsden, R. T. (2006). Cerebrospinal fluid leak rate after the use of BioGlue in translabyrinthine vestibular schwannoma surgery: A prospective study, *Neurotology*, **27**(1), 102–105.
5. Andreassen, T. T., and Jorgensen, P. H. (1985). Biomechanical properties and collagen formation in subcutaneously implanted cellulose sponges treated with fibrin sealant, *Eur. Surg. Res.*, **17**(4), 264–268.
6. Anema, J. G., Morey, A. F. Harris, R., MacPhee, M., and Cornum, R. L. (2001). Potential uses of absorbable fibrin bandage for genitourinary trauma, *World J. Surg.*, **25**, 1573–1577.
7. Arun, K., Maartens, N. F., and Kaye, A. H. (2003). Evaluation of the use of "BioGlue" in neurosurgical procedures, *J. Clin. Neurosci.*, **10**(6), 661–664.
8. Artzi, N., Shazly, T. M., Baker, A., B., Bon, A., and Edelman, E. R. (2009). Aldehyde-amine chemistry enables modulated biosealants with tissue-specific adhesion, *Adv. Mater.*, **21**(32–33), 3399–3403.
9. Baik, S. H., Kim, J. H., Cho, H. H., Park, S-N., Kim, Y. S., and Suh, H. (2010). Development and analysis of a collagen-based haemostatic adhesive, *J. Surg. Res.*, **164**, e221–e228.

10. BioGlue™, Instructions for Use Leaflet (2013).
11. Bitton, R., Josef, E., Shimshelashvili, I., Shapira, K., Seliktar, D., and Bianco-Peled, H. (2009). Phloroglucinol-based biomimetic adhesives for medical applications, *Acta Biomater.*, **5**, 1582–1587.
12. Bjork, J. W., Johnson, S. L., and Tranquillo, R. T. (2011). Ruthenium-catalyzed photo cross-linking of fibrin-based engineered tissue, *Biomaterials*, **32**, 2479–2488.
13. Brown, M. A., Daya, M. R., and Worley, J. A. (2009). Experience with chitosan dressings in a civilian EMS system, *J. Emerg. Med.*, **37**, 1–7.
14. Breated, J. H. (2006). *The Edwin Smith Surgical Papyrus: Hieroglyphic Transliteration, Translation and Commentary* (Kessinger Publishing, USA).
15. Busuttil, R. W. (2003). A comparison of antfibrinolytic agents used in hemostatic fibrin sealants, *J. Am. Coll. Surg.*, **197**, 1021–1028.
16. Chao, H.-H., and Torchiana, D. F. (2003). BioGlue: albumin/glutaraldehyde sealant in cardiac surgery, *J. Card. Surg.*, **18**, 500–503.
17. Cai, J., Yang, J., Du, Y. Fan, L., Qiu, Y. Li, J., and Kennedy, J. F. (2006). Enzymatic preparation of chitosan from the waste *Aspergillus niger* mycelium of a citric acid production plant, *Carbohyd. Polym.*, **64**(2), 151–157.
18. Celox Medical Website, accessed 01/06/2013, http://celoxmedical.com/#.
19. Chapman, W. C., Sherman, R., Boyce, S., Malawer, M., Hill, A., Buncke, G., Block, J., Fung, J. J., Clavien, P., Leef, K. F., Lebovic, G. S., Wren, S. M., Diethrich, E., and Goldstein, R. (2001). A novel collagen-composite offers effective haemostasis for multiple surgical indications: A result from a randomized controlled trial, *Surgery*, **129**, 445–450.
20. Chatterjee, S., Adhya, M., Guha, A. K., and Chatterjee, B. P. (2005). Chitosan from *Mucor rouxii*: Production an physiochemical characterization, *Process Biochem.*, **40**(1), 395–400.
21. Chen, T., Embree, H. D., Brown, E. M., Taylor, M. M., and Payne, G. F. (2003). Enzyme-catalyzed gel formation of gelatin and chitosan: potential for in situ applications, *Biomaterials*, **24**, 2831–2841.
22. Chen, T., Janjua, R., McDermott, M. K., Bernstein, S. L., Steidl, S. M., and Payne, G. F. (2006). Gelatin-based biomimetic tissue adhesive: Potential for retinal reattachment, *J. Biomed. Mater. Res. B Appl. Biomater.*, **77**, 416–422.
23. Chung, Y. S., Lee, K. E., Kim, H. Y., Choe J.-H., and Youn, Y.-K. (2010). Anti-adhesive effect and safety of sodium hyaluronate and sodium

carboxymethyl cellulose solution in thyroid surgery, *Asian J. Surg.*, **33**(1), 25–30.

24. Connolly, R. J. (2004). Application of the poly(-*N*-acetyl glucosamine)-derived rapid deployment hemostat trauma dressing in severe/lethal swine hemorrhage trauma models, *J. Trauma-Inj. Infec. Clin. Care*, **57**(1), S26–S28.

25. Cornum, R. L., Bell, J., Gresham, V., Brinkley, W., Beall, D., and MacPhee, M. (1999). Intraoperative use of absorbable fibrin adhesive bandage: Long term effects, *J. Urolog.*, **162**, 1817–1820.

26. CoStasis® Surgical Hemostat, Instructions for Use Leaflet.

27. Cunha, R. A., Soares, T. A., Rusu, V. H., Pontes, F. S. J., Franca, E. F., and Lins, R. D. The molecular structure and conformational dynamics of chitosan polymers: An integrated perspective from experiments and computational simulations, in: *The Complex World of Polysaccharides* (ed. Karunaratne, D. N.), InTech Publishers, Doi: 10.5772/2947 (2012).

28. Davidson, B. R., Burnett, S., Javed, M. S., Seifalian, A., Moore, D., and Doctor, N. (2000). Experimental study of a novel fibrin sealant for achieving haemostasis following partial hepatectomy, *Br. J. Surg.*, **87**(6), 790–795.

29. de la Portilla, F., Rada, R., León, E., Cisneros, N., Maldonado, V. H., and Esponosa, E. (2007). Evaluation of the use of BioGlue in the treatment of high anal fistulas: Preliminary results of a pilot study, *Dis. Colon Rectum*, **50**(2), 218–222.

30. Diabourov, M. (1991). Gelation: A review, *Polymer Int.*, **25**(3), 135–143.

31. Dowling, M. B., Kumar, R., Keibler, M. A., Hess, J. R. Bochicchio, G. V., and Raghavan, S. R. (2011). A self-assembling hydrophobically modified chitosan capable of reversible haemostatic action, *Biomaterials*, **32**(13), 3351–3357.

32. Elvin, C. M., Brownlee, A. G., Huson, M. G., Tebb, T. A., Kim, M., Lyons, R. E., Vuocolo, T., Liyou, N. E., Hughes, T. C., Edwards, G. A., Ramshaw, J. A. M., and Werkmeister, J. A. (2009). The development of photochemically crosslinked native fibrinogen as a rapidly formed and mechanically strong surgical sealant, *Biomaterials*, **30**, 2059–2060.

33. Elvin, C. M., Danon, S. J. Brownlee, A. G., White, J. F. Hickey, M., Liyou, N. E., Edwards, G. A., Ramshaw, J. A. M., and Werkmeister, J. A. (2010). Evaluation of photo-crosslinked fibrinogen as a rapid and strong tissue adhesive, *J. Biomed. Mater. Res. A*, **93**, 687–695.

34. Elvin, C. M., Vuocolo, T., Brownlee, A. G., Sando, L., Huson, M. G., Liyoi, N. E., Stockwell, P. R., Lyons, R. E., Kim, M., Edwards, G. A., Johnson, G., McFarland, G. A., Ramshaw, J. A. M., and Werkmeister, J. A. (2010). A highly elastic tissue sealant based on photopolymerised gelatine, *Biomaterials*, **31**, 8323–8331.
35. EN ISO 122442-1:2008 Standard.
36. EN ISO 22442-2:2008 Standard.
37. EN ISO 22442-3:2008 Standard.
38. Erdogan, D., and Van Gulik, T. M. (2008). The evolution of fibrinogen-coated collagen patch for use as a topic hemostatic agent, *J. Biomed. Mater. Res.* B, **85**, 272–278.
39. Ergenoglu, M. U., Yerebakan, H., and Kucukaksu, D. S. (2010). A new practical alternative for the control of sternal bleeding during cardiac surgery: Ankaferd blood stopper, *Heart Surg. For.*, **13**(6), E379–E380.
40. Evithrom®, Instructions for Use Leaflet (2013).
41. ExcelArresst®, Instructions for Use Leaflet (2010).
42. Fancy, D. A., and Kodadek, T. (1999). Chemistry for the analysis of protein-protein interactions: rapid and efficient cross-linking triggered by long wavelength light, *Proc. Natl. Acad. Sci. U. S. A.*, **96**, 6020–6024.
43. Farndale, R. W., Sixma, J. J., Barnes, M. J., and DeGroot, P. G. (2004). The role of collagen in thrombosis and haemostasis, *J. Thromb. Haemost.*, **2**, 561–573.
44. Ficke, J. R., and Pollak, A. N. (2007) Extremity war injuries: Development of clinical treatment principles, *J. Am. Acad. Orthop. Surg.*, **15**(10), 590–595.
45. Finley, D. S., Lee, D. I., Eichel, L., Uribe, C. A., McDougall, E. M., and Clayman, R. V. (2005). Fibrin glue-oxidized cellulose sandwich for laparoscopic wedge reconstruction of small renal lesions, *J. Urol.*, **173**(5), 1477–1481.
46. FloSeal® Surgical Hemostat, Instructions for Use Leaflet (2012).
47. Forer, B., Vasilyev, T., Brosh, T., Kariv, N., Trejo, L. L., Gil, Z., Katzir, A., and Fliss, D. M. (2006). Dural defect repair with fascia by a CO2 laser system in a porcine model, *Laryngoscope*, **116**(6), 1002–1006.
48. Foster, L. J. R., Thomson, K., Marcal, H., Butt, J., Watson, S., and Wakefield, D. (2010). A chitosan-vancomycin composite biomaterial as a laser activated surgical adhesive with regional antimicrobial activity, *Biomacromolecules*, **11**(12), 3563–3570.
49. Foster, L. J. R., and Butt, J. (2011). Chitosan films are NOT antimicrobial, *Biotech. Lett.*, **33**(2), 417–421.

50. Foster, L. J. R., and Karsten, E. A. (2012). A chitosan based, laser activated thin film surgical adhesive, "SurgiLux®": Preparation and demonstration, *J. Vis. Exp.*, **68**, e3527 DOI: 10.3791/3527.

51. Garcia, P., Mines, M. J., Bower, K. S., Hill. J., Menon, J., Tremblay, E., and Smith, B. (2009). Robotic laser tissue welding of sclera using chitosan films, *Lasers Surg. Med.*, **41**, 60–67.

52. Gardner, D. J., Oporto, G. S., Mills, R., and Samir, M. A. S. A. (2008). Adhesion and surface issues in cellulose and nanocellulose, *J. Adhes. Sci. Tech.*, **22**(5–6), 545–567.

53. Gelfoam® Plus, Instructions for Use Leaflet (2010).

54. Germanos, S., Bolanis, I., Saedon, M., and Baratsis, S. (2010), Control of presacral venous bleeding during rectal surgery, *Am. J. Surg.*, **200**(2), e33–e35.

55. Ghavamzadeh, R., Haddadi-Asl, V., and Mirzadeh, H. (2004). Bioadhesion and biocompatibility evaluations of gelatin and polyacrylic acid as a crosslinked hydrogel in vitro, *J. Biomater. Sci. Polym. Ed.*, **15**, 1019–1031.

56. Gregor, R. T. (2000). Management of chyle fistualizaton in association with neck dissection. *Otolaryngol. Head Neck Surg.*, **122**, 434–439.

57. Haines-Butterick, L., Rajagopal, K., Branco, M., Salick, D., Rughani, R., Pilarz, M., Lamm, M. S., Pochan, D. J., and Schneider, J. P. (2007). Controlling hydrogelation kinetics by peptide design for three-dimensional encapsulation and injectable delivery of cells, *Proc. Natl. Acad. Sci. U. S. A.*, **104**, 7791–7796.

58. Hanks, J. B., Kjaergard, H. K., and Hllingsbee, D. A. (2003). A comparison of the haemostatic effect of Vivostat® patient derived fibrin sealant with oxidised cellulose (Surgicel®) in multiple surgical procedures, *Eur. Surg. Res.*, **35**, 439–444.

59. Hattori, H., Amano, Y., Nogami, Y., Takaase, B., and Ishihara, M. (2010). Hemostasis for severe haemorrhage with photocrosslinkable chitosan hydrogel and calcium alginate, *Ann. Biomed. Eng.*, **38**(12), 3724–3732.

60. Haznedaroglu, B. Z., Beyazit, Y., Walker, S. L., and Haznedaroglu, I. C. (2012). Pleiotropic cellular, haemostatic, and biological actions of ankafed haemostat, *Crit. Rev. Oncol. Hemat.*, **83**(1), 21–34.

61. HemCon Medical Technologies Website, accessed 01/06/2013, http://www.hemcon.com/Products.aspx.

62. HemCon Medical Technologies Website, accessed 01/06/2013, http://www.hemcon.com/EducationCenter/AntibacterialAdvantages.aspx.

63. Hidas, G., Kastin, A., Mullerad, M., Shental, J., Moskovitz, B., and Nativ, O. (2006). Sutureless nephron-sparing surgery: use of albumin glutaraldehyde tissue adhesive (BioGlue), *Urology*, **67**(4), 697–700.

64. Higuchi, T. T., and Gettman, M. T. (2011). Essential Instruments in laparoscopic and robotic surgery, *Retrooper. Robot. Laparosc. Surg.*, **2011**, 9–22.

65. Hino, M., Ihiko, O., Honda, K.-I., Yamane, T., Ohta, K., Tabuko, T., and Tatsumi, N. (2000). Transmission of symptomatic parvovirus B19 infection by fibrin sealant used during surgery, *Br. J. Haematol.*, **108**, 194–195.

66. Holcomb, J. B., Pusateri, A. E., Harris, R. A., Reid, T. J. Beall, L. D., Hess, J. R., and MacPhee, M. J. (1999). Dry fibrin sealant dressings reduce blood loss, resuscitation volume, and improve survival in hypothermic coagulopathic swine with grade V liver injuries, *J. Trauma*, **47**, 233–240.

67. Holcomb, J. B., McClain, J. M., Pusateri, A. E., Beall, D., Macaitas, J. M., Harris, R. A., MacPhee, M. J., and Hess, J. R. (2000). Fibrin sealant foam sprayed directly on liver injuries decreases blood loss in resuscitated rats. *J. Trauma*, **49**, 246–250.

68. Hsu, S. H., and Lin, C. H. (2007). The properties of gelatin-poly (gamma-glutamic acid). hydrogels as biological glues, *Biorheology*, **44**, 17–28.

69. Jackson, M. R. (2001). Fibrin sealants in surgical practice: An overview, *Am. J. Surg.*, **182**(2), S1–S7.

70. Karsten, E., Watson, S. L., and Foster, L. J. R. (2012). Diversity of microbial species implicated in keratitis: A review, *The Open Ophthalmol. J.*, **6**, 117–131.

71. Khan, M. S., Bann, S. D., Darzi, A., and Butler, P. E. M. (2003). Use of suturing as a measure of technical competence, *Ann. Plast. Surg.*, **50**(3), 304–309.

72. Kim, S., Chung, E. H., Gilbert, M., and Healy, K. E. (2005). Synthetic MMP-13 degradable ECMs based on poly(*N*-isopropylacrylamide-co-acrylic acid) semi-interpenetrating polymer networks I: Degradation and cell migration, *J. Biomed. Mater. Res. A*, **75**, 73–88.

73. KiOmedine®, Brochure (2013).

74. Kirsch, A. J., Cooper, C. S., Gatti, J., Scherz, H. C., Canning, D. A., Zderic, S. A., and Snyder, H. M. (2001). Laser tissue soldering for hypospadias repair: results of a controlled prospective clinical trial, *J. Urol.*, **165**(2), 574–577.

75. Kjaergard, H. D., and Weis-Fogh, U. S. (1994). Important factors influencing the strength of autologous fibrin glue: The fibrin concentration and reaction time, *Eur. Surg. Res.*, **26**, 273–276.

76. Klimo, J., Khalil, A., Slotkin, J. R., Smith, E. R., Scott, R. M., and Goumnerova, L. (2007). Wound complications associated with the use of bovine serum albumin-glutaraldehyde surgical adhesive in pediatric patients, *Neurosurgery*, **60**(4), 305–309.

77. Kumar, U., and Albala, D. M. (2001). Newer techniques in intracorporeal tissue approximation: Suturing, tissue adhesives, and microclips, *Urol. Clin. N. Amer.*, **28**(1), 15–21.

78. Larson, M. J., Bowersox, J. C., Lim, R. C., and Hess, J. R. (1995). Efficacy of a fibrin haemostatic bandage in controlling haemorrhage from experimental arterial injuries, *Arch. Surg.*, **130**, 420–422.

79. Lauto, A., Foster, L. J., Ferris, L., Avolio, A., Zwaneveld, N., and Poole-Warren, L. A. (2004). Albumin-genipin solder for laser tissue repair, *Lasers Surg. Med.*, **35**(2), 140–145.

80. Lauto, A., Stoodley, M., Marcel, H., Avolio, A., Sarris, M., McKenzie, G., Sampson, D. D., and Foster, L. J. (2007). In vitro and in vivo tissue repair with laser-activated chitosan adhesive, *Laser. Surg. Med.*, **39**, 19–27.

81. Lauto, A., Mawad, D., and Foster, L. J. R. (2008). Adhesive biomaterials for tissue reconstruction, *Chem. Tech. Biotech.*, **83**(4), 464–472.

82. Lauto, A., Foster, L. J. R., Avolio, A., Sampson, D., Raston, C., Sarris, M., McKenzie, G., and Stoodley, M. (2008). Sutureless nerve repair with laser-activated chitosan adhesive: a pilot in vivo study, *J. Photomed. Laser Surg.*, **26**(3), 227–234.

83. Lauto, A., Foster, L. J., Sarris, M., McKenzie, G., and Stoodley, M. (2008). Sutureless nerve anastomosis with laser-activated chitosan adhesive, *Lasers Med. Sci.*, **23**(1), 104–105.

84. Lee, H., Lee, B. P., and Messersmith, P. B. (2007). A reversible wet/dry adhesive inspired by mussels and geckos, *Nature*, **448**, 338–341.

85. Lee, S., Pham, A. M., Pryor S. G., Tollefson, T., and Sykes, J. M. (2009). Efficacy of crosseal fibrin sealant (human) in rhytidectomy, *Arch. Facial Plast. Surg.*, **11**, 29–33.

86. LeMaire, S. A., Schmittling, Z. C., Coselli, J. S., Undar, A., Deady, B. A., Clubb, F. J., and Fraser, C. D. (2002). Bioglue surgical adhesive impairs aortic growth and causes anastomotic strictures, *Ann. Thorac. Surg.*, **73**(5), 1500–1506.

87. LeMaire, S. A., Carter, S. A., Won, T., Wang, X., Conklin, L. D., and Coselli, J. S. (2005). The threat of adhesive embolization: BioGlue leaks through needle holes in aortic tissue and prosthetic grafts, *Ann. Thorac. Surg.*, **80**(1), 106–111.

88. Li, C., Sajiki, T., Nakayama, Y. Fukui, M., and Matsuda, T. (2003). Novel visible-light induced photocurable tissue adhesive composed of multiply styrene-derivatized gelatin and poly(ethylene glycol) diacrylate, *J. Biomed. Mater. Res. B. Appl. Biomater.*, **66**(1), 439–446.

89. Liu, Y., Kopelman, D., Wu, L.-Q., Hijji, K., Attar, I., Priess-Bloom, O., and Payne, G. F. (2009). Biomimetic sealant based on gelatin and microbial transglutaminase: An initial in vivo investigation, *J. Biomed. Mater. Res. B*, **91**, 5–16.

90. Maitz, P. K., Trickett, R. I., Dekker, P., Tos, P., Dawes, J. M., Piper, J. A., Lanzetta, M., and Owen, E. R. (1999). Sutureless microvascular anastomoses by a biodegradable laser-activated solid protein solder, *Plast. Reconstr. Surg.*, **104**(6), 1726–1731.

91. Manju, S., Muraleedharan, C. V., Rajeev, A., Jayakrishnan, A., and Joseph, R. (2011). Evaluation of alginate dialdehde cross-linked gelatine hydrogel as a biodegradable sealant for polyester vascular graft, *J. Biomed. Mater Res. B Appl. Biomater.*, **98B**(1), 139–149.

92. Marine Polymer Technologies Inc., Syvek Website, accessed 01/06/2013, http://syvek.com/index.html.

93. Masayuki, I. (2002). Photocrosslinkable chitosan hydrogel as a wound dressing and a biological adhesive, *Trends Glycosci. Glycotechnol.*, **14**, 331–341.

94. Matsuda, S., Iwata, H., Se, N., and Ikada, Y. (1999). Bioadhesion of gelatin films crosslinked with glutaraldehyde, *J. Biomed. Mater. Res. A*, **45**(1), 20–27.

95. McCarthy, S. J., Gregory, K. W., Wiesmann, W. P., and Campbell, T. D. (2008). Wound dressing and method for controlling severe, life-threatening bleeding. US Patent 7,371,403B2, assigned to Providence Health System-Oregon; 2008.

96. McDermott, M. K., Chen, T., Williams, C. M., Markley, K. M., and Payne, G. F. (2004). Mechanical properties of biomimetic tissue adhesive based on the microbial transglutaminase-catalyzed crosslinking of gelatin, *Biomacromolecules*, **5**, 1270–1279.

97. MedMarket Dilligence LLC. (2012). Worldwide Surgical Sealants, Glues, Wound Closure and Anti-Adhesion Market, Forecast to 2017.

98. Menovsky, T., Beek, J. F., and van Gemert, M. J. (1996). Laser tissue welding of dura mater and peripheral nerves; a scanning electron microscopy study, *Lasers Surg. Med.*, **19**(2), 152–158.

99. Menovsky, T., and Beck, J. F. (2001). Laser, fibring glue, or suture repair of peripheral nerves: A comparative functional, histological and morphometric study in the rat sciatic nerve, *J. Neurosurg.*, **95**(4), 694–699.

100. Michaud, S. E., Wang, L. Z., Korde, N., Bucki, R., Randhawa, P. K., Pastore, J. J., Falet, H., Hoffmeister, K., Kuuse, R., Uibo, R., Herod, J., Sawyer, E., and Janmey, P. A. (2002). Purification of salmon thrombin and its potential as an alternative to mammalian thrombins in fibrin sealants, *Thromb. Res.*, **107**, 245–254.

101. Millesi, H. (1982). Peripheral nerve injuries: Nerve sutures and nerve grafting, *Scand. J. Plast. Reconstr. Surg. Suppl.*, **19**, 25–37.

102. Nageris, B. I., Zilker, Z., Zilker, M., Kariv, N., Feinmesser, R., Katzir, A. (2004). Esophageal incisions repair by CO_2 laser soldering. *Otolaryngol. Head Neck Surg.*, 131(6), 856–859.

103. Nakayama, Y., and Matsuda, T. (1999). Photocurable surgical tissue adhesive glues composed of photoreactive gelatin and poly(ethylene glycol) diacrylate, *J. Biomed. Mater. Res.*, **48**, 511–521.

104. No, H. K., and Meyers, S. P. (1995). Preparation and characterization of chitin and chitosan: A review, *J. Aquatic Food Prod. Tech.*, **4**(2), 27–52.

105. Okamoto, Y., Yano, R., Miyatake, K., Tomohiro, I., Shigemasa, Y., and Minami, S. (2003). Effects of chitin and chitosan on blood coagulation, *Carbohydr. Polym.*, **53**, 337–342.

106. Okino, H., Nakayama, Y., Tanaka, M., and Matsuda, T. (2002). In situ hydrogelation of photocurable gelatin and drug release, *J. Biomed. Mater. Res.*, **59**, 233–245.

107. Ono, K., Saito, Y., Yura, H., Ishikawa, K., Kurita, A., Akaike, T., and Ishihara, M. (2000). Photocrosslinkable chitosan as a biological adhesive, *J. Biomed. Mater. Res. A*, **49**(2), 289–295.

108. Oswald, A.-M., Joly, L.-M., Gury, C., Disdet, M., Leduc, V., and Kanny, G. (2003). Fatal intra-operative anaphylaxis related to aprotinin after local applications of fibrin glue, *Anesthesiology*, **99**, 762–763.

109. Oz, M. C., Rondinone, J. F., and Shargill, N. S. (2003). Floseal matrix: New generation topical haemostatic sealant, *J. Card. Surg.*, **18**, 486–493.

110. Ozbek, O., Acar, K., Osman, K., Saritas, K., Toy, H., Solak, Y., Ozbek, S., Kucukapan, A., Guler, I., Gaipov, A., Turk, S., Haznegaoglu, I. C. (2013). Short-term effects of Ankaferd haemostat for renal artery embolization: An experimental study, *Cardiovasc Intervent Radiol.*, **36**(2), 498–504.

111. Papov, V. V., Diamond, T. V., Biemann, K., and Waite, J. H. (1995). Hydroxyarginine-containing polyhenolic proteins in the adhesive plaques of the marine mussel *Mytilus edulis*, *J. Biol. Chem.*, **270**, 20183–20192.

112. Pecha, R. E., Prindiville, T., Kotfila, R., Ruebner, B., Cheung, A. T., and Trudeau, W. (1998). Gastrointestinal haemorrhage consequent to foreign body reaction to silk sutures: case series and review, *Gastrointest. Endoosc.*, **48**, 299–301.

113. Peng, H. T., and Shek, P. N. (2010). Novel wound sealants: Biomaterials and applications, *Expert Rev. Med. Devices*, **7**, 639–659.

114. Pochanavanich, P., and Suntomsuk, W. (2002). Fungal chitosan production and its characterization, *Lett. Appl. Microbiol.*, **35**(1), 17–21.

115. Poppas, D. P., Schlossberg, S. M., Richmond, H., Gilbert, D. A., and Devine, C. J. (1988). Laser welding in urethral surgery: Improved results with a protein solder, *J. Urol.*, **139**(2), 415–417.

116. Pusateri, A. E., McCarthy, S. J., Gregory. K. W., Harris, R. A., Cardenas, L., McManus, A. T., and Goodwin, C. W. (2003). Effect of a chitosan-based haemostatic dressing on blood loss and survival in a model of severe venous hemorrhage and hepatic injury in swine, *J. Trauma*, **54**, 177–182.

117. Raanani, E., Latter, D. A., Errett, L. E., Bonneau, D. B. Leclerc, Y., and Salasidis, G. C. (2001). Use of BioGlue in aortic surgical repair, *Ann. Thorac. Surg.*, **72**(2), 638–640.

118. Rao, S. B., and Sharma, C. P. (1997). Use of chitosan as a biomaterial: studies on its safety and haemostatic potential, *J. Biomed. Mater. Res.*, **34**, 21–28.

119. Rabea, E. I., Badawy, M. E. T., Stevens, C. V., Smagghe, G., and Steurbaut, W. (2003). Chitosan as an antimicrobial agent: applications and mode of action, *Biomacromolecules*, **4**(6), 1457–1465.

120. Recothrom®, Instructions for Use Leaflet (2010).

121. Reece, T. B., Maxey, T. S., and Kron, I. L. (2001). A prospectus on tissue adhesives, *Am J Surg.*, **182**, 40S–44S.

122. Renkins, K. L., Payner, T., Leipzig, T. J., Feuer, H., Morone, M. A., Koers, J. M. Lawson, K. J., Lentz, R., Shuey, H., Conaway, G. L., Andersson, G. B., An, H. S., Hickey, M., Rondinone, J. F. F., and Shargill, N. S. (2001). A multicentre, prospective, randomized trial evaluating a new haemostatic agent in spinal surgery, *Spine*, **26**, 1645–1650.

123. Reyes, J. M., Herretes, S., Pirouzmanesh, A., Wang, D. A., Elisseeff, J. H., Jun, A., McDonnell, P. J., Chuck, R. S., and Behrens, A. (2005). *Invest. Ophthalmol. Vis. Sci.*, **46**(4), 1247–1250.

124. Robertson, D., Lefebvre, G., Leyland, N., Wolfman, W., Allaire, C., Awadalla, A., Best, C., Contestabile, E., Dunn, S., Heywood, M., Leroux, N., Potestio, F., Rittenberg, D., Senikas, V., Soucy, R., and Singh, S. (2010). Adhesion prevention in gynaecological surgery; no. 243, June 2010, *Int. J. Gynecol. Obset.*, **111**(2), 193–197.

125. Rothwell, S. W., Fudge, J. M., Chen, W.-K., Reid, T. J., and Krishnamurti, C. (2002). Addition of a propyl gallate-based procoagulant to a fibrin bandage improved haemostatic performance in swine arterial bleeding model, *Thromb. Res.*, **108**(5–6), 335–340.

126. Rothwell, S. W., Fudge, J. M., Reid, T. J., and Krishnamurti, C. (2002) ε-Amino caproic acid additive decreases fibrin bandage performance in a swine arterial bleeding model, *Thromb. Res.*, **108**(5–6), 341–345.

127. Rothwell, S. W., Reid, T. J., Dorsey, J., Flournoy, W. S., Bodo, M., Janmey, P., and Sawyer, E. (2005). A salmon thrombin-fibrin bandage controls arterial bleeding in a swine aortotomy model, *J. Trauma*, **59**, 143–149.

128. Samuels, L. E., Shaw, P. M., and Blaum, L. C. (1996). Percutaneous technique for management of persistent airspace with prolonged air leak using fibrin glue, *Chest*, **109**, 1653–1655.

129. Sando, L., Danon, S., Brownlee, A. G., McCulloch, R. J., Ramshaw, J. A. M., Elvin, C. M., and Werkmeister, J. A. (2011). Photochemically crosslinked matrices of gelatine and fibrinogen promote rapid cell proliferation, *J. Tissue Eng. Regen. Med.*, **5**, 337–346.

130. Sawamura, Y., Asaoka, K., Terasaka, S., Tada, M., and Uchida, T. (1999). Evaluation of application techniques if fibrin sealant to prevent cerebrospinal fluid leakage: A new device for the application of aerosolized fibrin glue, *Neurosurgery*, **44**, 332–337.

131. Shahbazi, J., Marcal, H., Watson, S., Wakefield, D., Sarris, M., and Foster, L. J. R. (2011). Sutureless sealing of penetrating corneal wounds using a laser-activated thin film adhesive, *Laser Surg. Med.*, **43**, 490–498.

132. Shazly, T. M., Artzi, N., Boehning, F., and Edelman, E. R. (2008). Viscoelastic adhesive mechanics of aldehyde-mediated soft tissue sealants, *Biomaterials*, **29**(35), 4584–4591.

133. Shefy, N., Mintz, Y., Rivkind, A. I., and Shapira, S. C. (2006). Terror-related injuries: A comparison of gunshot wounds versus secondary-fragments-induced injuries from explosives, *J. Am. Coll. Surg.*, **203**, 297–303.

134. Shehtar-Harkavyk, I., and Bianco-Peled, H. (2004). On the relationship between adhesive properties and the structural features of fibrin sealants, *J. Adhes. Sci. Technol.*, **18**(12), 1415–1425.

135. Siber, J., Bukhari, T., Rudnicka, K., Murcin, L. A., and Tuthill, D. (2008). High-pressure fibrin sealant foam: an effective haemostatic agent for treating sever parenchymal hemorrhage, *J. Surg. Res.*, **144**, 145–150.

136. Sogias, I. A., Williams, A. C., and Khutoryanskiy, V. V. (2008). Why is chitosan mucoadhesive?, *Biomacromolecules*, **9**(7), 1837–1842.

137. Spotnitz, W. D., Dalton, M., S., Baker, J. W., and Nolan, S. P. (1987). Reduction of perioperative hemorrhage by anterior mediatinal spray application of fibrin glue during cardiac operations, *Ann. Thorac. Surg.*, **84**, 548–553.

138. Spotnitz, W. D. (2001). Commercial fibrin sealants in surgical care, *Am. J. Surg.*, **182**, 8S–14S.

139. Spotnitz, W. D., and Burks, S. (2008). Hemostats, sealants and adhesives: components of the surgical toolbox, *Transfusion*, **48**(7), 1502–1516.

140. Spotnitz, W. D., and Burks, S. (2010). Hemostats, sealants and adhesives II: update as well as how and when to use the components of the surgical toolbox, *Clin. App. Thromb. Hemost.*, **16**(5), 49–514.

141. Stilwell, R. G., Marks, M. G., Saferstein, L., and Wiseman, D. M. (1997). Oxidized cellulose: Chemistry, processing and medical applications, in *Handbook of Biodegradable Polymers* (ed., Domb, A. J., Kost, J., and Wiseman, D. M.), Chapter 15, Harwood Academic Publishers, Netherlands, pp. 291–306.

142. Sung, H. W., Huang, D. M., Chang, W. H., Huang, R. N., and Hsu, J. C. (1999). Evaluation of gelatin hydrogel crosslinked with various crosslinking agents as bioadhesives: in vitro study, *J. Biomed. Mater. Res.*, **46**, 520–530.

143. Surgiflo® Surgical Hemostat, Brochure and Instructions for Use Leaflet (2012).

144. Suzuki, S., and Ikada, Y. (2013). Sealing effects of cross-linked gelatin, *J. Biomater. Appl.*, **27**(7), 801–810.

145. Synaero™ Hemostatic Gel, Instructions for Use Leaflet (2013).

146. Taguchi, T., Saito, H., Uchida, Y., Sakane, M., Kobayashi, H., Kataoka, K., and Tanaka, J. (2004). Bonding of soft tissues using a novel tissue adhesive consisting of a citric acid derivative and collagen, *Mater. Sci. Eng. C*, **24**, 775–780.

147. Taguchi, T., Saito, H., Aoki, H., Uchida, Y., Sakane, M., Kobayashi, H., and Tanaka, J. (2006). Biocompatible high-strength glue consisting of citric acid derivative and collagen, *Mater. Sci. Eng. C*, **26**, 9–13.

148. Thrombin-JMI®, Instructions for Use Leaflet (2013).

149. Tisseel® (2012) Instructions sheet.

150. Yu, M., Hwang, J., and Deming, T. J. (1999). Role of L-3,4-dihydroxyl-phenylalanine in mussel adhesive proteins, *J. Am. Chem. Soc.*, **121**, 5825–5826.

151. Van der Meijden, P. E. J., Munnix, I. C. A., Auger, J. M., Govers-Riemslag, J. W. P., Cosemans, J. M. E. M., Kuijpers, M. J. E., Spronk, H. M., Watson, S. P., Renné, T., and Hememskerk, J. W. M. (2009). Dual role of collagen in factor XIII-dependent formation, *Blood*, **114**(4), 881–890.

152. Wang, L. Z., Gorlin, J., Michaud, S. E., Janmey, P. A., Goddeau, R. P., Kuuse, R., Uibo, R., Adams, D., and Sawyer, E. S. (2000). Purification of salmon clotting factors and their use as tissue sealants, *Thromb. Res.*, **100**, 537–548.

153. Wax, M. K., Ramadan, H. H., Ortiz, O., and Wetmore, S. J. (1997). Contemporary management of cerebrospinal fluid rhinorrhea, *Otolaryngol. Head Neck Surg.*, **116**, 442–449.

154. Wedmore, I., McManus, J. G., Pusateri, A. E., and Holcomb, J. B. (2006). A special report on the chitosan-based hemostatic dressing: Experience in current combat operations, *J. Trauma*, **60**, 655–658.

155. Woodworth, B. A., Chandra, R. K., LeBenger, J. D., Ilie, B., and Schlosser, R. J. (2009). A gelatin-thrombin matrix for haemostasis after endoscopic sinus surgery, *Am. J. Otolaryn. Head Neck Med. Surg.*, **30**, 49–53.

Chapter 10

Bioadhesive Systems for Drug Delivery

Ryan F. Donnelly and A. David Woolfson

School of Pharmacy, Queen's University Belfast,
Medical Biology Centre, 97 Lisburn Road, Belfast BT9 7BL, UK

r.donnelly@qub.ac.uk

Bioadhesion is commonly defined as the adhesion between two materials, at least one of which is a biological in nature. Over the past few decades, bioadhesion-enhanced drug delivery has received a great deal of attention. Bioadhesive dosage forms may be designed to enable prolonged retention at the site of application, providing a controlled rate of drug release for improved therapeutic outcome. Application of dosage forms to mucosal surfaces may be of benefit to drug molecules not amenable to the oral route, such as those that undergo acid degradation or extensive first pass metabolism. The bioadhesive ability of a dosage form is dependent upon a variety of factors, including the nature of the mucosal tissue, and the physicochemical properties of the polymeric formulation. This chapter aims to provide an overview of the various aspects various aspects of bioadhesion, bioadhesive materials, factors affecting bioadhesion, evaluating methods, and finally various bioadhesive drug delivery systems (buccal, nasal, ocular, gastro, vaginal and rectal).

Bioadhesion and Biomimetics: From Nature to Applications
Edited by Havazelet Bianco-Peled and Maya Davidovich-Pinhas

ISBN 978-981-4463-98-0 (Hardcover), 978-981-4463-99-7 (eBook)
www.panstanford.com

10.1 Introduction

Mucus membranes (mucosae) are the moist surfaces lining the walls of various body cavities such as the gastrointestinal and respiratory tracts [1,2]. They consist of a connective tissue layer (the lamina propria) above which is an epithelial layer, the surface of which is made moist usually by the presence of a mucus layer. The epithelia may be either single layered (e.g. the stomach, small and large intestines and bronchi) or multilayered/stratified (e.g. in the oesophagus, vagina and cornea). The former contain goblet cells which secrete mucus directly onto the epithelial surfaces; the latter contain, or are adjacent to tissues containing, specialized glands such as salivary glands that secrete mucus onto the epithelial surface [1–3]. Mucus is present either as a gel layer adherent to the mucosal surface or as a luminal soluble or suspended form. The major components of all mucus gels are mucin glycoproteins, lipids, inorganic salts and water, the latter accounting for more than 95% of their weight, making them a highly hydrated system [1–3]. The major functions of mucus are that of protection and lubrication. While bioadhesive drug delivery systems adhere to biological substrates in general, mucoadhesion refers to a special case, where the delivery system adheres to the mucous overlying a mucus membrane. This chapter provides a full overview of bio- and mucoadhesive systems for drug delivery purposes. Such systems are typically used for sustained administration of drug substances across biological barriers. However, depending on the site of administration, rapid delivery is also possible. Application of such systems to the skin, oral, vaginal, rectal or nasal mucosa avoids hepatic first pass metabolism, instead delivering the therapeutic agent directly to the systemic circulation, thus enhancing bioavailability. By also side-stepping the harsh low-pH environment of the stomach and gastrointestinal enzymes and bypassing the low permeability of the small intestine, enhanced delivery of biotherapeutics, such as peptides and proteins is also possible. Since mucosal sites are replete with immune-presenting cells, vaccination efficiency can be improved. A detailed description of bioadhesion and mucoadhesion, the theories of mucoadhesion and the factors affecting mucoadhesion are covered in detail elsewhere in this book. In this chapter, formulation of bio- and mucoadhesive dosage forms is covered in detail, as are the properties of the polymers that are the basis of such systems. Also described are the techniques used for the determination of mucoadhesion and the most common routes of mucoadhesive administration.

10.2 Routes of Administration for Bio- and Mucoadhesive Drug Delivery Systems

The mucus membrane or mucosa is the moist tissue that lines organs and body cavities such as the mouth, gut, rectum, genital area, nose and eye lid. The anatomical differences of mucus membrane are outlined in Table 10.1.

Table 10.1 Anatomical differences of the mucus membrane

Mucus membrane	Relevant anatomical features
Buccal [2]	Buccal mucosa surface area approximately 30 cm^2
	Comprising three distinct layers: epithelium, basement membrane, and connective tissues
	Buccal mucosa (Fig. 10.1), sublingual and soft palate comprise non-keratinized tissue. Gingival tissue and hard palate are keratinized tissues
	Thickness of buccal epithelium is in the range of 500–800 μm, 40–50 cells thick
	Mucus secreted by salivary glands, as a component of saliva, forming a 0.1–0.7 mm thick layer
	Turnover time for buccal epithelium 5–6 days
	Permeability barrier property of oral mucosa due to intercellular materials derived from membrane-coating granules
Nasal [3]	Nasal cavity surface area 160 cm^2
	Lined with mucous membrane containing columnar cells, goblet cells, and basal cells
	Columnar cells are covered with cilia, apart from the anterior part of the nasal cavity
	Both keratinized and non-keratinized epithelial cells present depending upon location within nasal cavity
	Cilia responsible for mucociliary clearance
	Mucus secreted by the submucosal glands and the goblet cells, forming a mucus layer approximately 5–20 μm thick
	Nasal cavity length approximately 60 mm
	Nasal cavity volume approximately 20 mL
	Turn-over time for mucus is usually 10–15 min

(*Continued*)

Table 9.1 (*Continued*)

Mucus membrane	Relevant anatomical features
Ocular [4]	Cornea is composed of five layers: epithelium, Bowman's layer, stroma, Descemet's membrane, and endothelium Epithelium consists of 5–6 layers of cells, with the cells of the basal layer being columnar, and the outermost cells flattened polygonal cells Tight junctions present between the basal cells of the corneal epithelium At the corneal margin, the conjunctiva is structurally continuous with the corneal epithelium The conjunctival tissue is permeable to molecules up to 20,000 Da, whereas the cornea is impermeable to molecules greater than 5000 Da The conjunctiva contains around 1.5 million goblet cells, which synthesize secretory mucins and peptides A volume of about 2–3 μL of mucus is secreted daily A turnover of the mucus layer occurs in approximately 15–20 h Exposed part of the eye is covered by a thin fluid layer: precorneal tear film
Vaginal [5]	Tear film thickness is approximately 3–10 μm Length of vagina varies from 6 to 10 cm The epithelial layer consists of the lamina propia and stratified squamous epithelium A cell turnover of about 10–15 layers is estimated to be in the order of 7 days Although there are no glands in the vaginal mucosa, the surface is usually covered with vaginal fluid Major components of vaginal fluid are cervical mucus and vaginal fluid from the well-vascularised mucosa The volume, viscosity, and pH of the cervical mucus vary with age and during the menstrual cycle
Rectal [6]	Length approximately 15–20 cm Surface area of approximately 300 cm^2 Epithelium consists of a single layer of cylindrical cells and goblet cells secreting mucus Flat surface, without villi and with three major folds, the rectal valves Approximately 3 mL of mucus with a neutral pH spread over the surface

Source: Adapted from Shaikh et al. [1].

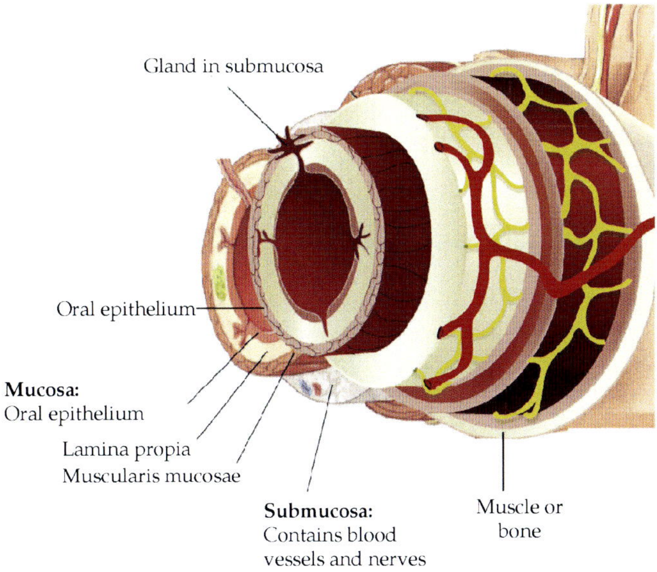

Figure 10.1 Cross section of the oral mucosa. Adapted with permission from McCrudden et al. [81].

Examples of mucoadhesive drug delivery systems which have been formulated include powders, compacts, sprays, semisolids and films. Developments in delivery systems include compacts for drug delivery into the oral cavity [7], powders and nanoparticles for delivery to nasal mucosa [8,9] and oral strips for the tongue and buccal cavity [10]. More advanced and alternative mucoadhesive formulations have been suggested, including buccal films which may provide a greater degree of flexibility and comfort than adhesive tablets. They may also improve the residence time compared to oral gels [11]. The most commonly-used polymers for such formulations are cellulose derivatives, poly (acrylic acids) such as Carbopol® and Gantrez® copolymers, such as poly(methylvinylether/maleic anhydride) [12].

10.3 Mucoadhesive Polymers

Polymers for bioadhesive/mucoadhesive drug delivery systems must have certain characteristics [13–16]. The polymer and its degradation products should be non-toxic, non-absorbable and non-irritant. The polymer should form strong bonds with the mucus or epithelial cell surface and should adhere quickly to moist tissue. It should possess site specificity, allow easy incorporation of the drug and offer no hindrance to drug release. The polymer must resist decomposition on storage or during the shelf life of the dosage form. However, the delivery system should be capable of trauma-free removal from the substrate. Generally, this will not be an issue if adhesion is to the mucus layer, which is regularly shed [1,2]. Table 10.2 presents some of the most commonly used polymers for mucoadhesive purposes with their chemical structures.

Table 10.2 Chemical structures of some bioadhesive polymers used in drug delivery

Chemical name (abbreviation)	Chemical structure
Poly(ethylene glycol) (PEG)	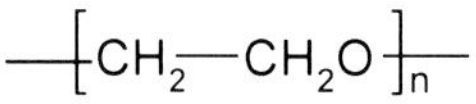
Poly(vinyl alcohol) (PVA)	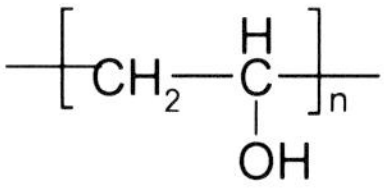
Poly(vinyl pyrrolidone) (PVP)	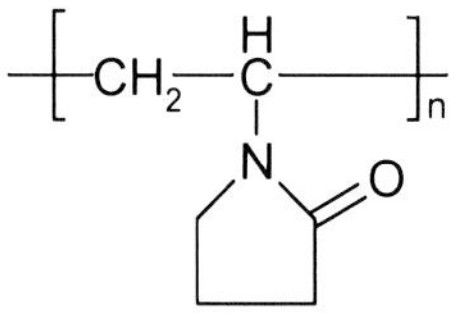
Poly(acrylic acid) (PAA or Carbopol®)	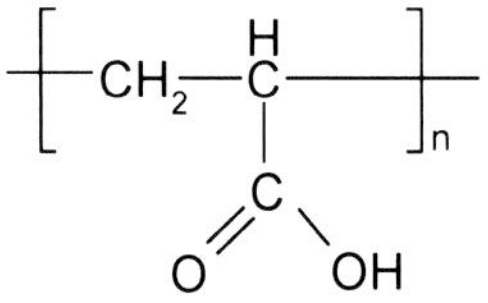
Poly(hydroxyethyl methacrylate) (PHEMA)	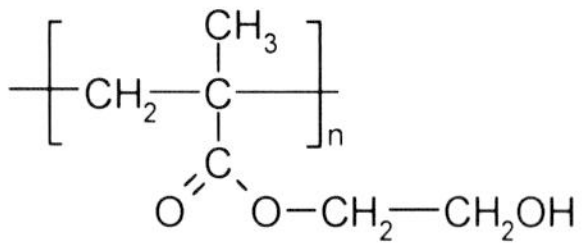
Chitosan	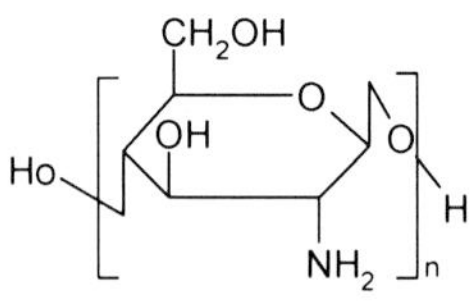

Hydroxyethylcellulose (HEC): R = H and CH_2CH_2 OH Hydroxypropylcellulose (HPC): R = H and $CH_2CH(OH)CH_3$ Hydroxypropylmethylcellulose (HPMC): R = H, CH_3 and $CH_2CH(OH)CH_3$ Methylcellulose: R = H and CH_3 Sodium carboxymethylcellulose (NaCMC): R = H and CH_2COONa	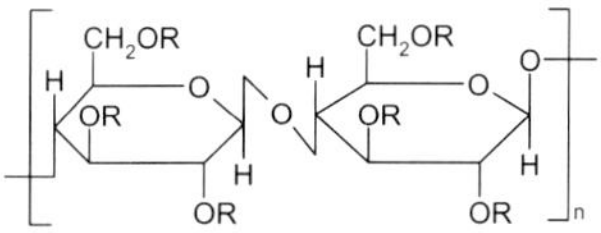
Acrylated polymers [84] Acrylate monomers used to form acrylate polymers are based on the structure of acrylic acid, which consists of a vinyl group and a carboxylic acid terminus. Other typical acrylate monomers are derivatives of acrylic acid, such as methyl methacrylate in which one vinyl hydrogen and the carboxylic acid hydrogen are both replaced by methyl groups, and acrylonitrile in which the carboxylic acid group is replaced by the related nitrile group.	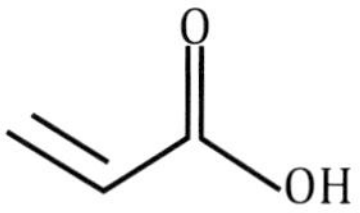 Acrylic acid 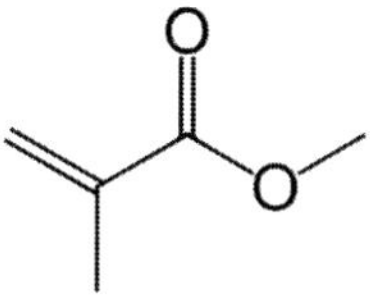 Methylmethacrylate 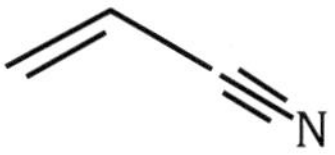 Acrylonitrile
Thiomers [83,85] Thiolated polymers or "thiomers" are mucoadhesive polymers which display thiol group bearing side chains. Low molecular mass thiol group-bearing compounds are covalently bound to a polymeric backbone. This backbone typically consists of a well-established polymer with a long track record of safe use, such as chitosan or poly(acrylic acid). Thiomers generally exhibit mucoadhesive, permeation enhancing, controlled release as well as enzyme and efflux pump inhibitory properties.	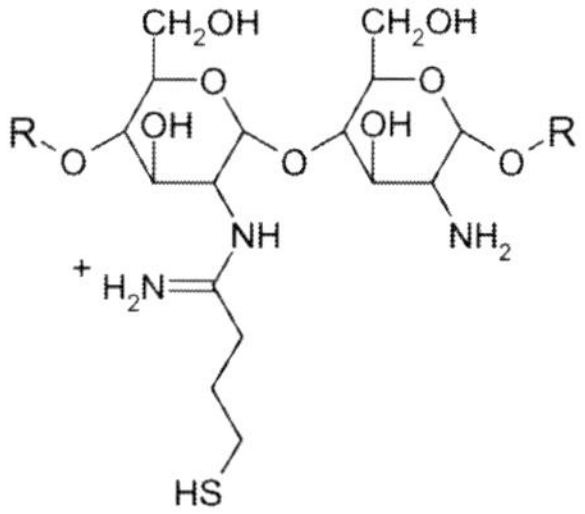 Thiolated chitosan derivative

Source: Adapted from Shaikh et al. [1].

There are a number of broad categories of polymers which possess the suitable characteristics for adherence to biological surfaces. The first category is polymers that adhere through non-specific, non-covalent interactions which are primarily electrostatic in nature. The second is polymers possessing hydrophilic functional groups which hydrogen bond with similar groups on biological substrates. The third group is polymers that bind to specific receptor sites on the cell or mucus surface. Examples of polymers in the third category include lectins. Lectins have the ability to bind sugars selectively in a non-covalent manner. They are proteins or glycoprotein complexes of non-immune origin [17]. Lectins have been used for drug targeting applications due to their ability to attach themselves to carbohydrates on the mucus or epithelial cell surface [18,19]. The fourth group contains thiolated polymers, which are hydrophilic macromolecules exhibiting free thiol groups on the polymeric backbone. Thiolated polymers have many advantageous mucoadhesive properties, including improved tensile strength, rapid swelling and water uptake behaviour. These improved properties are due to the presence of thiol groups in the polymer. They allow the formation of stable covalent bonds with the cystein subdomains of mucus glycoproteins, which results in increased residence time and improved bioavailability [20].

10.4 Factors Affecting Mucoadhesion

Mucoadhesion is affected by a number of factors, including hydrophilicity, molecular weight, cross-linking, swelling, pH, and the concentration of the active polymer [13,16,20].

10.4.1 Hydrophilicity

A hydrophilic molecule is one that has the tendency to interact or be dissolved by water and other polar substances. Bioadhesive polymers possess numerous hydrophilic functional groups. These hydrophilic functional groups (i.e. hydroxyl and carboxyl groups) allow hydrogen bonding with the substrate and swelling in aqueous media. These occurrences allow maximal exposure of potential anchor sites. The stronger the hydrogen bonding, the stronger the adhesion, in general [21]. Polymers which have swollen have the maximum distance between their chains, which leads to increased chain flexibility and increases the efficiency of substrate penetration [1].

10.4.2 Molecular Weight

Polymer molecular weight is important as it determines many physical properties. The optimum molecular weight for maximum mucoadhesion depends on the type of polymer. Interpenetration of polymer molecules is favoured by low-molecular-weight polymers and entanglements are favoured at high molecular weights. Generally, bioadhesive forces increases with the molecular weight of the polymer up to 100,000. Above this, there is no further gain [86].

10.4.3 Cross-Linking and Swelling

Polymers consist of repeating structural subunits (monomers) connected by covalent bonds. In addition to the bonds which hold monomers together in a polymer chain, many polymers form bonds between neighbouring chains. These bonds can be formed directly between the neighbouring chains, or two chains may bond to a third common molecule. This phenomenon is called cross-linking. Although not as strong or rigid as the bonds within the chain, these cross-links have an important effect on the polymer. Cross-linking density is inversely proportional to the degree of swelling [22]. Low cross-linking density results in higher flexibility and hydration rate. Also, increasing the surface area of the polymer improves mucoadhesion. A lightly cross-linked polymer is most suitable to obtain a high degree of cross-linking. However, an excess of moisture will result in a degree of cross-linking that is too great and the resulting slippery mucilage can be easily removed from the substrate [23]. The addition of adhesion promoters such as free polymer chains and polymers grafted onto the performed network can improve the mucoadhesion of cross-linked polymers [20].

10.4.4 Spatial Conformation

The spatial conformation of the polymer (the shape which the molecule preferentially occupies in a certain medium) is also extremely important. An example of the importance of spatial conformation to bioadhesion can be seen when comparing the molecular weight of dextrans (19,500,000) to that of polyethylene glycol (PEG) (200,000). Although the molecular weight of dextran is much higher, both polymers have a similar adhesive strength. This phenomenon is due to the helical conformation of dextran which may shield many adhesively active groups, unlike PEG which have a linear conformation [1,2].

10.4.5 pH

The pH at the bioadhesive to substrate interface can influence the adhesion of polymers that contain ionisable groups. For example, many polymers used for bioadhesive purposes in drug delivery are polyanions possessing carboxylic acid functionalities. The approximate pK_a for many of these polymers is pH 4–5. The maximum adhesive strength of these polymers is between 4 and 5 and decreases gradually above pH 6. It has been proven that the protonated carboxyl groups, rather than the ionized carboxyl groups, react with the mucin molecules. It has been suggested that the mechanism of this action is the simultaneous formation of many hydrogen bonds [24].

10.4.6 Concentration of Active Polymer

It has been reported by Ahuja [16], that an optimal concentration of corresponding to optimal mucoadhesion exists, in liquid mucoadhesive formulations. In highly concentrated systems, mucoadhesion drops above this optimal concentration. This has been explained by fact that in concentrated solutions, coiled molecules become solvent poor and the chains available for interpenetration are sparse. For solid dosage forms such as tablets, Duchene et al. [14] showed that the higher the polymer concentration the stronger the mucoadhesion.

10.4.7 Drug/Excipient Concentration

The ratio of drug to excipient may affect the mucoadhesive effects. For example, the effect of propranol hydrochloride content on the adhesive properties of Carbopol® hydrogels was investigated by Blanco-Fuente et al. [25]. Increased adhesion was noted when water was limited in the system. This was due to an increase in elasticity, caused by complex formation between the drug and polymer. In the presence of large amounts of water, the complex precipitated out which led to a decrease in the adhesive character. Increasing toluidine blue O (TBO) concentration in mucoadhesive patches based on poly (methylvinylether/maleic acid) significantly increased mucoadhesion to porcine cheek tissue [26]. It was suggested that this was due to increased internal cohesion within the patches, caused by electrostatic interactions between the cationic drug and anionic polymer.

10.4.8 Other Factors Affecting Mucoadhesion

The initial force of application may affect mucoadhesion [27]. High bioadhesive strength and enhances interpenetration are induced by higher forces [28]. The initial contact time also influences mucoadhesion. The greater the initial contact time between the bioadhesive and the substrate, the greater the swelling and interpenetration of polymer chains [29]. Physiological variables such as the rate of mucus turnover can affect mucoadhesion [30]. The mucoadhesive surface can vary depending on the site in the body and the presence of local or systemic disease; this in turn will affect mucoadhesion [29].

10.5 Delivery Systems

As detailed above, a range of different strategies and different dosage form types have been employed for bio- and mucoadhesive delivery. The dosage form type necessarily depends upon the route of administration and the barrier properties presented at the various sites of application, as well as the amount of moisture present at the site.

10.5.1 Oral Mucoadhesive Delivery Systems

The accessibility of mucoadhesive delivery systems to the oral mucosa has resulted in its significant interest as a delivery route. The most commonly used routes are the buccal and sublingual. The non-keratinized epithelium in the oral cavity, i.e. the soft palate, the mouth floor, the ventral side of the tongue and the buccal mucosa offer a relatively permeable barrier for drug transport [31]. Oral mucosae are composed of multiple layers of cells, which show various patterns of differentiation. The functions of the different regions in the oral cavity affect the patterns of differentiation [32]. Blood supply to the oral cavity tissue is provided by the external carotid artery. This artery branches to the maxillary lingual and facial artery. The oral mucosae do not contain mucus secreting globlet cells; however, mucins are found in human saliva. Saliva consists of 99% water and the mucins dissolved within it form a gel of 20–100 μm thickness [32]. Paracellular transport is generally followed by hydrophilic compounds and large or highly polar molecules whereas lipophilic drugs follow transcellular transport through the lipid bilayer [33]. Advantages of the oral route over others include the bypassing of the hepatic first pass metabolism, improvement in drug bioavailability, improved patient compliance, excellent accessibility, unidirectional drug flux and improved barrier permeability that intact skin [34,35]. The application of mucoadhesive delivery systems to the oral cavity allows both local and systemic delivery. Local therapy is use to treat conditions such as aphthous ulceration gingivitis, periodontal disease and xerostoma. Common dosage forms include adhesive gels, tablets, films, patches, ointments, mouthwashes and pastes.

The most frequently used dosage form for buccal drug delivery up to now has been adhesive tablets (Fig. 10.2a). Tablets can be applied to many regions of the oral cavity, such as cheeks, lips, gums and the palate. Such tablets allow the drinking eating and speaking without any major discomfort. A number of recent studies have examined the suitability of adhesive tablets as a oral mucoadhesive delivery system. Perioli studied the influence of compression force on tablet behaviour and drug release rate for mucoadhesive buccal tablets. Tablets were prepared from hydroxyethyl cellulose (HEC) and Carbopol 940 in a 1:1 ratio at varying compression forces. It was found that mucoadhesion performance and drug release were affected by compression force as increasing compression force decreased the in vitro and in vivo drug release profiles. An increased compression force also resulted in the best in vivo mucoadhesive and hydration time. Tablets prepared at the highest forces caused pain during in vivo application and had to be detached by human volunteers [36]. Another study carried out by Shermer et al. [37] evaluated the efficacy and tolerability of a mucoadhesive patch compared with a pain relieving oral solution for the treatment of aphthous stomatitis. The mucoadhesive patch was more successful than the oral solution in terms of healing time and pain intensity after 12 and 24 h. After 1 h of treatment, the local adverse effects were significantly less frequent with the mucoadhesive patch patients as compared with the oral solution patients. Donnelly et al. reported on a mucoadhesive patch (Fig. 10.2b) containing TBO as a potential system for use in photodynamic antimicrobial chemotherapy (PACT) of oropharyngeal candidiasis. The patches were prepared from aqueous blends of poly (methyl vinyl ether/maleic anhydride) and tripropyleneglycol methyl ether. It was found that short application times of TBO loaded patches allowed the treatment of recently acquired oropharyngeal candidiasis, if they are caused solely by planktonic cells. For biofilms, it was suggested that longer application times would be needed [26].

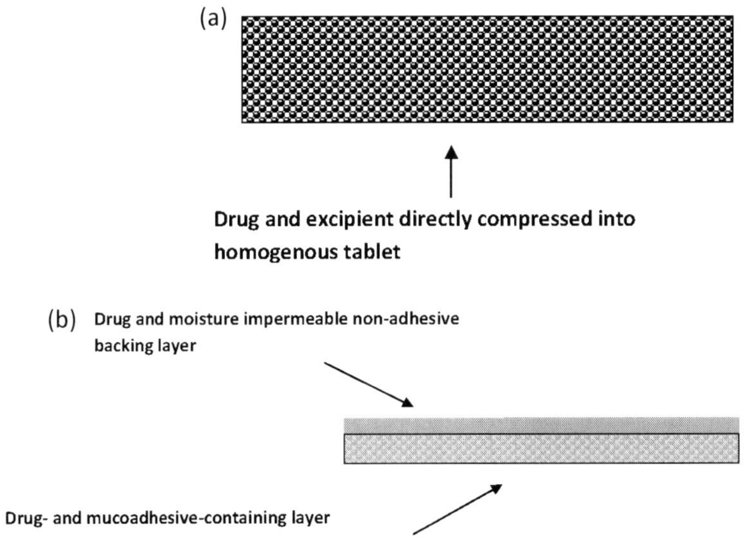

Figure 10.2 Homogenous mucoadhesive tablet (a) and bilayer mucoadhesive patch (b).

Desai developed oral mucoadhesive patch formulations for the delivery of fenretinide for site-specific chemoprevention of oral cancer [38]. In vitro and in vivo release studies were investigated. The solubilisation of fenretinide in simulated saliva was studied by incorporating non-ionic surfactants (Tween® 20 and 80, and Brij® 35 and 98), bile salts (sodium salt of cholic, taurocholic, glycocholic, and deoxycholic acids), phospholipid (lecithin), and novel polymeric solubilizer (Souplus®). Adhesive and drug release layers were prepared using solvent casting. The adhesive layers were prepared from Hydroxypropyl methylcellulose and the drug release layers were prepared from Eudragit® RL PO with or without solubilizers with incorporated fenretinide. The oral mucoadhesive patches were formed by attaching drug and adhesive layers onto a backing layer (Tegaderm™ film). The study found that solubilizer-free patch exhibited poor in vitro and in vivo controlled drug release. The incorporation of single or mixed solubilizers in fenretinide/Eudragit® patches significantly improved continuous in vitro/in vivo fenretinide release. The paper concluded that fenretinide/Eudragit® RL PO patches with 20 wt% Tween® 80 + 40 wt% sodium deoxycholate solubilizers exhibit excellent release behaviour for further preclinical and/or clinical evaluation in oral cancer chemoprevention.

A number of mucoadhesive buccal delivery formulations are currently being commercialised or undergoing clinical trials. Generex Biotechnology developed a novel liquid aerosol formulation (Oralin®) which allowed precise insulin dose delivery via a metered dose inhaler in the form of fine aerosilized droplets directed into the mouth [39]. The oral aerosol formulation is rapidly absorbed through the buccal mucosal epithelium and it provides the plasma insulin levels necessary to control postpranial glucose rise in diabetic patients. Advantages of this novel delivery system include rapid absorption, a user-friendly administration technique, precise dosing control and bolus delivery. However, this has yet to reach the market. Products that have reached the market include BioAlliance Pharma's miconazole tablet (Lauriad®), and Aphtach® which are triamcinolone acetonide buccal tablets from Teijin Ltd.

10.5.2 Nasal Mucoadhesive Delivery Systems

The human nasal mucosa is a highly dense vasculature network and is a relatively permeable membrane structure. The area of the nasal mucosa is normally approximately 150 cm^2 [40]. The nasal epithelium exhibits high permeability as only two cell layers separate the nasal lumen from the dense blood vessel network in the lamina propria. The main lining of the nasal cavity is the respiratory epithelium. It allows the clearance of mucus by the mucociliary system and is composed of ciliated and non-ciliated columnar cells, globlet cells and basal cells. The respiratory epithelium is covered by a mucus layer which can be divided into the periciliary layer and a gel-like upper layer. The periciliary layer is less viscous than the gel-like layer. Mucociliary clearance allows the removal of foreign substances and particles from the nasal cavity, therefore preventing them from reaching the upper airways. This process is facilitated by the cilia which propel the mucous layer towards the nasopharynx [41]. Advantages of the nasal route of delivery include rapid uptake and the avoidance of first pass hepatic metabolism. Disadvantages include local toxicity and irritation, mucociliary clearance of 5 min, the presence of proteolytic enzymes and the possible influence of pathological conditions (cold and allergies). The application of bioadhesive delivery systems such as liquids, semisolids and solids may significantly increase retention time.

The delivery of proteins and peptides via the nasal route can be compromised by the brief residence time at the mucosal surface. Some bioadhesive polymers have been suggested to extend residence time and improve protein uptake across the nasal mucosa. A recent study by McInnes et al. used gamma scintigraphy to quantify nasal residence time of bioadhesive formulations [42]. This study also investigated the absorption of insulin. A conventional nasal spray was compared with three lyophilized nasal insert formulations (1–3% w/w hydroxypropylmethyl cellulose, HPMC) in a four-way crossover study with six male volunteers. It was found that the 2% w/w HPMC lyophilized formulation was most successful at achieving an extended residence time. This concentration had the ability to adhere to the nasal mucosa rapidly without overhydration.

Viscosity enhancing mucosal delivery systems for the induction of an adaptive immune response against a viral antigen was examined by Coucke [43]. Spray dried formulations of starch (Amioca®) and poly (acrylic acid) (Carbopol® 974P) were used as carriers of the vial antigen. An in vivo rabbit model was used to compare these formulations for intranasal delivery of heat-inactivated influenza virus combined with LTR 192G adjuvant. The formulations tested successfully induced a systemic anti-HA antibody response after intranasal vaccination with a whole virus influenza vaccine.

The development, characterization and safety aspects of nasal drug products consisting of functionalized mucoadhesive polymers (i.e. polycarbophil, hyaluronan and amberlite resin) has been examined recently. Work into the use of mucosal vaccines for the induction of a systemic immune response has recently been carried out and it has been found that the addition of a mucoadhesive polymer to the vaccine formulation increases the affinity for the mucus membrane and enhances the stability of the preparation. Examples of such vaccines include influenza vaccines, diphtheria vaccines and tetanus [44].

A recent pilot study into the use of a nasal morphine-chitosan formulation for the treatment of pain in cancer patients, concluded that the systems was well tolerated and accepted by patients and that it could lead to a rapid onset of pain relief [45]. A recent study by Tzachev et al. compared a commercially available decongestant solution which was non-mucoadhesive, with a mucoadhesive solution of xylometazoline in human patients with perennial allergic rhinitis. The study found that the mucoadhesive formulation produces a more prolonged clinical effect than the commercially available produced. A study by Luppi et al. investigated the preparation of albumin nanoparticles carrying cyclodextrins for the nasal delivery of the anti-Alzheimer drug tacrine. In this instance, the nasal route was selected as peroral administration of this drug is associated in low bioavailability. Bovine serum albumin nanoparticles were prepared using a coacervation method followed by thermal cross-linking. The nanoparticles were then loaded by soaking from solutions of tacrine hydrochloride and lyophilizing. The authors concluded that the study indicate that albumin nanoparticles carrying native and hydrophilic derivatives can be employed for the formulation of mucoadhesive nasal formulations [46].

10.5.3 Ocular Mucoadhesive Delivery Systems

The ocular route is mainly used for local treatment of eye pathologies. Conventional delivery methods to the eye are generally unsuccessful due to the inherent protective mechanisms of the eye (tear production, tear flow and blinking), the limited area of absorption and the lipophilic character of the corneal epithelium [47]. The precorneal tear film is the first structure encountered by an ocular dosage form. It consists of three distinct layers. The outer layer is of oily and lipid nature and prevents tear evaporation. The middle layer contains an aqueous salt solution and the inner layer is a mucus layer which is secreted by the conjunctiva goblet cells and lachrymal gland. This layer maintains moisture in the corneal and conjunctival epithelia. The ocular membranes comprise the cornea which is non-vascularized and the conjunctiva which is vascularized. The major pathway for ocular drug penetration is considered to be the corneal epithelium. It consists of five or six layers of non-keratinized squamous cells [48]. Solutions and suspensions are swiftly washed from the cornea and ointments can alter the tear refractive index and blur the vision. Therefore, prolonging the residence time by mucoadhesion may provide the required conditions of successful ocular delivery.

Recently, Sensoy et al. prepared bioadhesive sulfacetamide sodium microspheres to increase residence time on the ocular surface, thereby attempting to enhance treatment efficacy of ocular keratitis [49]. Polymer mixtures containing pectin, polycarbophil and HPMC at different ratios were used to prepare microparticle formulations via a spray drying method. An in vivo model consisting of New Zealand rats with keratitis caused by *Pseudomonas aeruginosa* and *Staphylococcus aureus* was used in the study. An optimized formulation containing sulfacetamide sodium-loaded polycarbophil microspheres at a polymer drug ratio of 2:1 was found to be most suitable for an ocular application. Advances in gene transfer technology have allowed its application to the treatment of several chronic diseases the affect the ocular surface. The efficacy and mechanism of action of a bioadhesive DNA nanocarrier prepared from hyaluronan (HA) and chitosan (CS) was investigated by De la Fuente et al. [50]. The formulation was specifically designed for topical ophthalmic gene therapy. An in vivo study using rabbits found that the nanoparticles entered the corneal and conjunctival epithelial cells and were assimilated by the cells. The nanoparticles also provided efficient delivery of the associated plasmid DNA inside the cells reaching significant transfection levels. Work into reducing the movement of ocular films across the eye with the addition of mucoadhesive polymers has been examined by Alza [44]. It has been found that films have a tendency to move across the eye without the addition of mucoadhesive polymers. The improvements associated with the addition of mucoadhesive polymers have been found to minimise ocular irritation and burning sensations.

A recent clinical study was conducted by Baeyens et al. [51] into the use of soluble bioadhesive ophthalmic drug inserts (BODI®) in dogs with external ophthalmic diseases (conjunctivitis, superficial corneal ulcers or keratoconjunctivitis sicca). These results were compared to classical Tiacil® eye drops from Virbac Laboratories. The BODI® inserts reduced the treatment to a single application, therefore improving patient compliance. A number of mucoadhesive polymers have been incorporated into ophthalmic gels to increase gel efficacy. Examples include NyoGel® (timolol, Novartis) and Pilogel® (pilocarpine hydrochloride, Alcon Laboratories) [52].

10.5.4 Vaginal Mucoadhesive Delivery Systems

The vagina is a fibrovascular tube connecting the uterus to the outer surface of the body. The vaginal epithelium consists of a stratified epithelium and lamina propia. The vagina offers asubstantial area for drug absorption. The numerous folds in the epithelium increase its total surface area. A rich vascular network surrounds the vagina, whereas the vaginal epithelium is covered by a film of moisture consisting mainly of cervical mucus and fluid secreted from the vaginal wall. The dosage forms that are usually used for the vaginal route are solutions, gels, suspensions, suppositories, creams and tablets. They all have a short residence time [53–55]. Bioadhesives may control drug release and extend the residence time of such formulations. They may contain drug or act in conjunction with moisturizing agents.

Recent advances in polymeric technology have increased the potential of vaginal gels. Vaginal gels are semisolid polymeric matrices comprising small amounts of solid, dispersed in relatively large amounts of liquid that have been used in systems for microbiocides, contraceptives, labour inducers and other substances. An acid-buffering bioadhesive vaginal clotrimazole (antifungal) and metronidazole (antoprotozoal and antibacterial) tablets for the treatment of genitourinary tract infections were developed by Alam [56]. Poly(carbophil) and sodium carboxymethyl cellulose were found to be a good combination for an acid-buffering bioadhesive vaginal tablet in release studies. It was also found that these bioadhesive polymers held the tablet for over 24 h inside the vagina. The in vitro spreadability of a commercially available conventional tablet (Infa-V®) was compared to the aforementioned formulation and it was found to be comparable to the commercially available gel. It was also found that the acid-buffering bioadhesive tablet produced better antimicrobial action than marketed intravaginal drug delivery systems (Infa-V®, Candid-V® and Canesten® 1). Gel formulations consisting of clomiphene citrate (CLM) for the local treatment of human papilloma virus infections were prepared by Cevher [57]. Formulations were prepared, including the polyacrylic acid (PAA) polymers such as Carbopol® 934P (C934P), Carbopol® 971P (C971P), Carbopol® 974P (C974P) in their various concentrations and their conjugates containing thiol groups. The gels which contained C934P-Cys showed the highest adhesiveness and mucoadhesion and increasing polymer concentration significantly decreased drug release.

A number of clinical trials are underway on microbiocidal gels. Microbiocidal gels are intended to improve mucosal permeation rate of microbiocides for the prevention of sexually transmitted diseases. Currently, a 1% tenofovir gel is being investigated in phage 2 clinical trials for determining the safety and acceptability of vaginal microbiocides [58]. Clinical trials for contraceptive gels such as Buffergel® is in phases 2 and 3 clinical trial comparing it to Gynol 2® marketed product [58]. The Prostin E2® suppository containing dinoprostone found that administration of prostaglandin E2 gel was more effective in inducing labour when tested in clinical trials [98]. Another phage 3 clinical trial conducted by Janssen Pharmaceutica of a mucoadhesive system based on intraconazole vaginal cream containing cyclodextrins and other ingredients. The clinical tests found that the cream was well tolerated and was an effective delivery system for selected vaginal delivery [59].

10.5.5 Rectal Mucoadhesive Delivery Systems

The rectum is a part of the colon. It is 10 cm in length and has a surface area of 300 cm^2. The main function of the rectum is the removal of water. The rectum has a relatively small surface area for drug absorption. The absorption of drugs through the rectum is generally achieved by a simple diffusion process through the lipid membrane. Advantages of the rectal route of delivery include the avoidance of first pass metabolism. The use of bioadhesive delivery systems in the rectum can also decrease the drug migration distance.

A study by Kim et al. [60] developed a thermoreversible flurbiprofen liquid suppository base composed of a poloxamer and sodium alginate for the improvement of the rectal bioavailability of flurbiprofen. Cyclodextrin derivatives enhanced the aqueous solubility of flurbiprofen. An in vivo study in rats showed that flurbiprofen liquid suppository containing the cyclodextrin derivative HP-beta-CD showed excellent bioavailability. It was concluded that HP-beta-CD could successfully be used as a solubility enhancer in liquid suppositories containing poorly water-soluble drugs.

10.5.6 Cervical and Vulval Delivery Systems

A number of recent studies have been carried out to investigate application of mucoadhesive delivery systems to the cervix and vulva due to the increasing prevalence of human papilloma virus associated neoplasias. A novel bioadhesive cervical patch containing 5-fluorouracil for the treatment of cervical intraepithelial neoplasia (CIN) was developed by Woolfson et al. The patch was a drug-loaded bioadhesive patch prepared from a gel containing 2% w/w Corbopol® 981 plasticized with 1% w/w glycerine. The casting solvent used was ethanol:water 30:70. The film was bonded directly to a backing layer formed from thermally cured poly(vinyl chloride) emulsion. Substantial drug release through human cervical tissue samples was observed over approximately 20 h [61,62]. However, despite promising clinical trial data, this formulation has not been roprietaryed. This is as a direct result of the successful use of the cervical cancer vaccine and the efficacy of laser-based cold coagulation treatments in the developed world. Undoubtedly, the bioadhesive patch could be useful in resource-poor countries, though substantial charity funding would be required for additional safety and efficacy trials. The design, physiochemical characterization and clinical evaluation of bioadhesive delivery systems for photodynamic therapy (PDT) of difficult to manage vulval neoplasias and dysplasias was carried out by Donnelly et al.

[63]. Aminolevulic acid (ALA) is commonly delivered to the vulva using creams or solutions and covered by an occlusive dressing. These types of formulations are poor at staying in place. Therefore, the production of a bioadhesive patch was suggested as an alternative delivery system. The patches were shown to release more ALA over 6 h than the proprietary cream (Porphin®, 20% w/w ALA). Clinically, the patch was extensively used in the successful PDT of vulval intraepithelial neoplasia, lichen sclerosis, squamous hyperplasia, Paget's disease and vulvodynia and the technology has now been licensed to Swedish Pharma AB for further clinical investigation and commercial development.

10.5.7 Gastrointestinal Mucoadhesive Delivery Systems

The oral route is undoubtedly the most favoured route of administration. It represents the most convenient route of drug administration, being characterized by high patient compliance. The mucosal epithelium along the gastrointestinal tract varies. In the stomach, the surface epithelium consists of a single layer of columnar cells whose apical membrane is covered by a conspicuous glycocalyx. A thick layer of mucus covers the surface to protect against aggressive luminal content. The small intestine in characterized by an enormous surface area available for the absorption of nutrients and drugs. The intestinal epithelium consists of a single layer of three types of columnar cells: enterocytes, globlet cells and enteroendocrine cells. The large intestine (colon) has the same cell populations as the small intestine, and its main function is the absorption of water and electrolytes. The role of mucus in the intestine is to facilitate the passage of food along the intestinal tract and to protect the gut from bacterial infections [64]. Problems associated with the oral route include hepatic first pass metabolism, degradation of drug during absorption, mucus covering GI epithelia and high turnover of mucus covering GI epithelia and the high turnover of mucus. Recently, the gastrointestinal tract (GIT) delivery has emerged as a very important route of administration. Bioadhesive retentive system involves the use of bioadhesive polymers, which can adhere to the epithelial surface in the GIT. The use of bioadhesive systems would increase GI transit time and increase bioavailability.

Gastric retention formulations (GRFs) made of naturally occurring carbohydrate polymers and containing riboflavin in vitro were studied by Ahmed et al. [65]. The swelling and dissolution characteristics of the formulations were examined in vitro and the in vivo behaviour in fasting dogs for gastric retention was also investigated. The bioavailability of riboflavin from the GRFs was studied in fasted healthy humans and compared to an immediate release formulation. It was found that on immersion in gastric juice, the GRFs swelled rapidly and released their drug load in zero order fashion over a period of 24 h. The GRF stayed in the stomach of fasted dogs for up to 9 h, then disintegrated and reached the colon in 24 h. Also, the bioavailability of riboflavin from a large size GRF was more than triple of that measured after administration of an immediate release formulation.

Nanoparticle carriers with bioadhesive properties were evaluated for their adjuvant potential for oral vaccination by Salma et al. [66]. Thiamine was used as a specific ligand–nanoparticle conjugate (TNP) to target specific sites within the gastrointestinal tract, namely enterocytes and Pweyer's patches. The affinity of the nanoparticles to the gut mucosa was studied in orally inoculated rats. The thiamine-coated nanoparticles showed promise as a particulate vector for oral vaccination and immunotherapy.

10.6 Methods of Determining Mucoadhesion

The evaluation of bioadhesive and mucoadhesive properties is fundamental to the development of novel bioadhesive delivery systems. Mucoadhesive formulations are usually in either hydrogel or film form.

10.6.1 Determination of Mucoadhesion in Polymer Gel Systems

The past decade has seen a range of methods employed to study mucoadhesion phenomena and mucoadhesive properties of hydrogel systems. Table 10.3 lists some of the methods employed and how they are carried out. These methods specifically pertain to hydrogel systems.

Table 10.3 Methods of studying mucoadhesion.

Direct assays	
Tensiometry	Force required to dislodge two surfaces, one coated with mucus, the other solid dosage form consisting of mucoadhesive hydrogel
Flow through	Flow rate dV/dt required to dislodge two surfaces; useful for microparticulate dosage forms
Colloidal gold staining [82]	An in vitro method called "mucin-gold staining" developed for the quantitative comparison of mucoadhesive properties of various hydrogels. The technique employs red colloidal gold particles which are stabilized by the adsorbed mucin molecules (mucin-gold conjugates). Upon interaction with mucin-gold conjugates, mucoadhesive hydrogels develop a red colour on the surface. Thus, the mucoadhesive properties of hydrogels can be compared quantitatively by measuring the intensity of the red colour.
In vivo techniques	Endoscopy, gamma scintigraphy
Molecular mucin-based assays	
Viscometry and rheology	Intrinsic viscosity [*Z*] can be related to complex size via MHKS (Mark–Houwink–Kuhn–Sakurada) a coefficient
Dynamic light scattering	Diffusion coefficient, D, can be related to complex size via MHKS coefficient
Turbidity, light scattering	SEC MALLS (size exclusion chromatography—multi-angle laser light scattering) particularly useful for determining MW of mucin, turbidity, semiquantitative indicator
Analytical ultracentrifugation	Change in MW (sedimentation equilibrium), sedimentation coefficient ratio of complex to mucin
Surface plasmon resonance imaging methods	Needs mobile and immobile phase, atomic force microscopy (conventional and gold labelled), scanning tunnelling microscopy

Source: Adapted from Junginer, Verhoef and Thanou [67].

Study of the flow and deformation of polymer gel formulations (i.e. their rheological characteristics) may be useful in predicting the mucoadhesive ability of a polymeric formulation. Hassan and Gallo [68] first suggested this approach by determining the rheological interaction between a polymer gel and a mucin solution. This experiment proved that a polymer gel and mucin solution mixture evoked a larger rheological response than the sum of the values of polymer and mucin. Varying rheological results may be attributed to differences in mucin type and concentration [69,70] and in polymer concentrations [70,71]. It has been recommended that rheological methods should not be used as a standalone method for determining mucoadhesive properties of polymer gels [72].

There is a much higher proportion of in vitro mucoadhesion studies compared to in vivo due to the time, cost and ethical constraints of in vivo studies. However, some work into in vivo mucoadhesive testing has been completed recently, specifically in monitoring mucoadhesion on tissue surfaces such as GIT or the buccal cavity. A study by Chung focussed on the in vivo transit time for bioadhesive beads in the rat GIT [73]. A ^{51}Cr-labelled bioadhesive were inserted at selected time intervals, GIT was removed and cut into sections and radioactivity was measured. Davis completed a non-invasive study to determine the transit of the mucoadhesive agent. The formulation used contained a gamma-emitting radionuclide and the release characteristics and polymer position was examined by gamma scintigraphy [74]. MRI has also been recently used to detect the time and location of the release of a mucoadhesive formulation using dry Gd-DOTA powder [75].

10.6.2 Determination of Mucoadhesion in Film Systems

It has been recently pointed out that no standard apparatus is available for testing bioadhesive strength and therefore, an inevitable lack of uniformity between test methods has arisen. However, there are three main methods for testing bioadhesive films [1]. The main testing methods are tensile/shear tests and peel strength.

10.6.3 Tensile and Shear Tests

The most common technique for the determination of force of separation in bioadhesive testing is the application of a force perpendicularly to the tissue/adhesive interface. During this process, a state of tensile stress is set up. During the shear stress, the direction of the forces is re-orientated so that it acts along a joint interface. In both tensile and shear modes, an equal pressure is distributed over the contact area [76]. A recent literature review has found that the tensile strength method is the most common technique used for bioadhesion testing. A commercially available texture profile analyzer (TA-XT Plus, stable microsystems, Surrey, UK) operating in bioadhesive test mode can be used to measure thee force required to remove bioadhesive films from excised tissue in vitro [23,77–80]. Figure 10.3 illustrates the texture profile analyzer in bioadhesion mode.

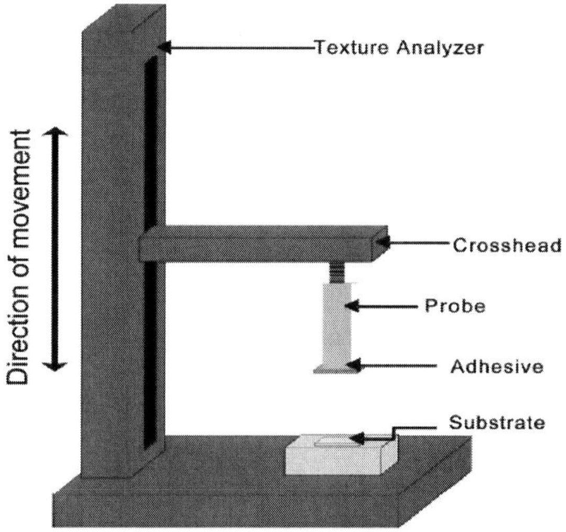

Figure 10.3 Texture Profile Analyser in bioadhesion test mode.

10.6.4 Peel Strength

The peel strength is based on the calculation of energy required to detach the patch from the substrate. It is of most use when the bioadhesive system is formulated as a patch [77]. The texture analyser (as illustrated in Fig. 10.4) operating in tensile test mode and coupled with a sliding lower platform can also be used to determine peel strength of such formulations [77]. Figure 10.4 illustrates a typical set-up used to determine the peel strength of bioadhesive films.

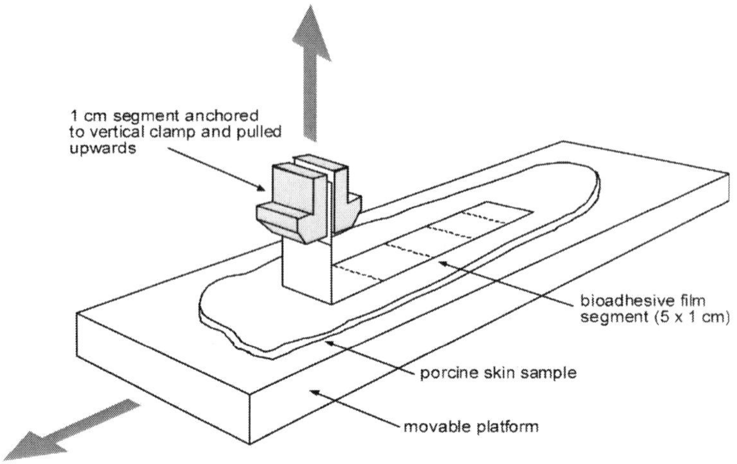

Figure 10.4 Simplified representation of the test set-up used to determine peel strength of bioadhesive films. Reproduced with permission from McCarron et al. [77].

10.7 Conclusion

Bio- and mucoadhesive dosage forms offer prolonged contact at the site of administration, low enzymatic activity and enhanced patient compliance. The formulation of mucoadhesive drug delivery systems depends on the selection of suitable polymer with excellent mucosal adhesive properties and biocompatibility. Researchers are now looking beyond traditional polymers and focussing on next-generation mucoadhesive polymers such as lectins and thiols. These polymers offer greater attachment and retention of dosage forms. However, these novel mucoadhesive formulations will require much more work before they are marketed in approved products for human use.

References

1. Shaikh R, Raj Singh TR, Garland MJ, Woolfson AD, Donnelly RF. Mucoadhesive drug delivery systems. *Journal of Pharmacy & Bioallied Sciences.* 2011; **3**(1): 89–100.
2. Salamat-Miller N, Chittchang M, Johnston TP. The use of mucoadhesive polymers in buccal drug delivery. *Advanced Drug Delivery Reviews.* 2005; **57**(11): 1666–1691.

3. Ugwoke MI, Agu RU, Verbeke N, Kinget R. Nasal mucoadhesive drug delivery: Background, applications, trends and future perspectives. *Advanced Drug Delivery Reviews*. 2005; **57**(11): 1640–1665.
4. Ludwig A. The use of mucoadhesive polymers in ocular drug delivery. *Advanced Drug Delivery Reviews*. 2005; **57**(11): 1595–1639.
5. Valenta C. The use of mucoadhesive polymers in vaginal delivery. *Advanced Drug Delivery Reviews*. 2005 Nov [cited August 18, 2011]; **57**(11): 1692–1712.
6. Edsman K, Hägerström H. Pharmaceutical applications of mucoadhesion for the non-oral routes. *The Journal of Pharmacy and Pharmacology*. 2005; **57**(1): 3–22.
7. Ponchel G, Touchard F, Duchêne D, Peppas NA. Bioadhesive analysis of controlled-release systems. I. Fracture and interpenetration analysis in poly(acrylic acid)-containing systems. *Journal of Controlled Release*. 1987; **5**(2): 129–141.
8. Nagai T, Konishi R. Buccal/gingival drug delivery systems. *Journal of Controlled Release*. 1987; **6**(1): 353–360.
9. Sayin B, Somavarapu S, Li XW, Thanou M, Sesardic D, Alpar HO, et al. Mono-N-carboxymethyl chitosan (MCC) and N-trimethyl chitosan (TMC) nanoparticles for non-invasive vaccine delivery. *International Journal of Pharmaceutics*. 2008; **363**(1–2): 139–148.
10. Ilango R. In-vitro studies on buccal strips of glibenclamide using chitosan. *Indian Journal of Pharmaceutical Sciences*. 1997; **59**(5): 232.
11. Anders R, Merkle HP. Evaluation of laminated muco-adhesive patches for buccal drug delivery. *International Journal of Pharmaceutics*. 1989; **49**(3): 231–240.
12. Woolfson A. Development and characterisation of a moisture-activated bioadhesive drug delivery system for percutaneous local anaesthesia. *International Journal of Pharmaceutics*. 1998; **169**(1): 83–94.
13. Jiménez-castellanos MR, Zia H, Rhodes CT. Mucoadhesive drug delivery systems. *Drug Development and Industrial Pharmacy*. 1993; **19**(1–2): 143–194.
14. Duchěne D, Touchard F, Peppas NA. Pharmaceutical and medical aspects of bioadhesive systems for drug administration. *Drug Development and Industrial Pharmacy*. 1988; **14**(2–3): 283–318.
15. Reinhart CT, Peppas NA. Solute diffusion in swollen membranes. Part II. Influence of crosslinking on diffusive properties. *Journal of Membrane Science*. 1984; **18**: 227–239.

16. Ahuja A, Khar RK, Ali J. Mucoadhesive drug delivery systems. *Drug Development and Industrial Pharmacy*. 1997; **23**(5): 489–515.
17. Smart JD. An in vitro assessment of some mucosa-adhesive dosage forms. *International Journal of Pharmaceutics*. 1991; **73**(1): 69–74.
18. Naisbett B, Woodley J. The potential use of tomato lectin for oral drug delivery. 1. Lectin binding to rat small intestine in vitro. *International Journal of Pharmaceutics*. 1994; **107**(3): 223–230.
19. Nicholls T. Lectins in ocular drug delivery: An investigation of lectin binding sites on the corneal and conjunctival surfaces. *International Journal of Pharmaceutics*. 1996; **138**(2): 175–183.
20. Peppas NA, Little MD, Huang Y. Bioadhesive controlled release systems. In *Handbook of Pharmaceutical Controlled Release Technology* (Wise DL, ed), New York, Marcel Dekker; 2000. p. 255–269.
21. Roy S, Pal K, Anis A, Pramanik K, Prabhakar B. Polymers in mucoadhesive drug delivery system: A brief note. *Designed Monomers and Polymers*. 2009; (12): 483–495.
22. Gudeman LF, Peppas NA. Preparation and characterization of pH-sensitive, interpenetrating networks of poly(vinyl alcohol) and poly(acrylic acid). *Journal of Applied Polymer Science*. 1995; **55**(6): 919–928.
23. McCarron PA, Woolfson AD, Donnelly RF, Andrews GP, Zawislak A, Price JH. Influence of plasticizer type and storage conditions on properties of poly(methyl vinyl ether-co-maleic anhydride) bioadhesive films. *Journal of Applied Polymer Science*. 2004; **91**(3): 1576–1589.
24. Park H, Robinson JR. Physico-chemical properties of water insoluble polymers important to mucin/epithelial adhesion. *Journal of Controlled Release*. 1985; **2**: 47–57.
25. Blanco-Fuente H. In-vitro bioadhesion of carbopol hydrogels. *International Journal of Pharmaceutics*. 1996; **142**(2): 169–174.
26. Donnelly RF, McCarron PA, Tunney MM, David Woolfson A. Potential of photodynamic therapy in treatment of fungal infections of the mouth. Design and characterisation of a mucoadhesive patch containing toluidine blue O. *Journal of Photochemistry and Photobiology. B, Biology*. 2007; **86**(1): 59–69.
27. Smart JD. An in vitro assessment of some mucosa-adhesive dosage forms. *International Journal of Pharmaceutics*. 1991; 73(1): 69–74.
28. Peppas NA, Buri PA. Surface, interfacial and molecular aspects of polymer bioadhesion on soft tissues. *Journal of Controlled Release*. 1985; **2**: 257–275.

29. Kamath KR, Park K. Mucosal adhesive preparation. In *Encyclopedia of Pharmaceutical Technology* (Swarbrick J, Boylan JC, ed), New York, Marcel Dekker; 1992. P. 133.

30. Lehr C-M, Poelma FGJ, Junginger HE, Tukker JJ. An estimate of turnover time of intestinal mucus gel layer in the rat in situ loop. *International Journal of Pharmaceutics*. 1991; **70**(3): 235–240.

31. Leung SHS, Robinson JA. Polyanionic polymers in bioadhesive and mucoadhesive drug delivery. *ACS Symposium Series*. 1992; **480**: 269–284.

32. Squier CA, Wertz PW. Structure and function of the oral mucosa and implications for drug delivery. *Drugs and the Pharmaceutical Sciences*. 1996; **74**: 1–26.

33. Shojaei AH, Berner B, Li X. Transbuccal delivery of acyclovir: I. In vitro determination of routes of buccal transport. *Pharmaceutical Research*. 1998; **15**(8): 1182–1188.

34. al-Achi A, Greenwood R. Buccal administration of human insulin in streptozocin-diabetic rats. *Research Communications in Chemical Pathology and Pharmacology*. 1993; **82**(3): 297–306.

35. Bouckaert S, Remon J-P. In-vitro bioadhesion of a buccal, miconazole slow-release tablet. *Journal of Pharmacy and Pharmacology*. 1993; **45**(6): 504–507.

36. Perioli L, Ambrogi V, Giovagnoli S, Blasi P, Mancini A, Ricci M, et al. Influence of compression force on the behavior of mucoadhesive buccal tablets. *AAPS PharmSciTech*. 2008; **9**(1): 274–281.

37. Shermer A, Amicji B, Trau H, Nathansohn N, Mizrahi B, Domb, AJ. Efficacy of a mucoadhesive patch compared with an oral solution for treatment of aphthous stomatitis. *Drugs R. D.* 2008; **9**(1): 29–35.

38. Desai K-GH, Mallery SR, Holpuch AS, Schwendeman SP. Development and in vitro-in vivo evaluation of fenretinide-loaded oral mucoadhesive patches for site-specific chemoprevention of oral cancer. *Pharmaceutical Research*. 2011; **28**(10): 2599–2609.

39. Modi P, Mihic M, Lewin A. The evolving role of oral insulin in the treatment of diabetes using a novel RapidMist System. *Diabetes/Metabolism Research and Reviews*. 2002; **18**(1): S38–S42.

40. Nagai T, Machida Y. Bioadhesive dosage forms for nasal administration, In *Bioadhesive Drug Delivery Systems*. (Lenaerts V, Gurney R, ed), Florida, Boca Raton: CRC Press; 1990.

41. Marttin E, Schipper NGM, Verhoef JC, Merkus FWHM. Nasal mucociliary clearance as a factor in nasal drug delivery. *Advanced Drug Delivery Reviews*. 1998; **29**(1–2): 13–38.
42. McInnes FJ, O'Mahony B, Lindsay B, Band J, Wilson CG, Hodges LA, et al. Nasal residence of insulin containing lyophilised nasal insert formulations, using gamma scintigraphy. *European Journal of Pharmaceutical Sciences*. 2007; **31**(1): 25–31.
43. Coucke D, Schotsaert M, Libert C, Pringels E, Vervaet C, Foreman P, et al. Spray-dried powders of starch and crosslinked poly(acrylic acid) as carriers for nasal delivery of inactivated influenza vaccine. *Vaccine*. 2009; **27**(8): 1279–1286.
44. Kharenko EA, Larionova NI, Demina NB. Mucoadhesive drug delivery systems (Review). *Pharmaceutical Chemistry Journal*. 2009; **43**(4): 200–208.
45. Pavis H, Wilcock A, Edgecombe J, Carr D, Manderson C, Church A, et al. Pilot study of nasal morphine-chitosan for the relief of breakthrough pain in patients with cancer. *Journal of Pain and Symptom Management*. 2002; **24**(6): 598–602.
46. Luppi B, Bigucci F, Corace G, Delucca A, Cerchiara T, Sorrenti M, et al. Albumin nanoparticles carrying cyclodextrins for nasal delivery of the anti-Alzheimer drug tacrine. *European Journal of Pharmaceutical Sciences*. 2011; **44**(4): 559–565.
47. Saettone MF, Burgalassi S, Chetoni P. Ocular bioadhesive drug delivery systems. In *Bioadhesive Drug Delivery Systems* (Mathiowitz E, Chickering DE, III, Lehr C-M, eds), Marcel Dekker, Inc., New York, 1999; 601–640.
48. Joshi, A. Microparticulates for ophthalmic drug delivery. *Journal Ocular Pharmacology*. 1994; **10**: 29–45.
49. Sensoy D, Cevher E, Sarici A, Yilmaz M, Ozdamar A, Bergişadi N. Bioadhesive sulfacetamide sodium microspheres: Evaluation of their effectiveness in the treatment of bacterial keratitis caused by Staphylococcus aureus and Pseudomonas aeruginosa in a rabbit model. *European Journal of Pharmaceutics and Biopharmaceutics*. 2009; **72**(3): 487–495.
50. de la Fuente M, Seijo B, Alonso MJ. Bioadhesive hyaluronan-chitosan nanoparticles can transport genes across the ocular mucosa and transfect ocular tissue. *Gene Therapy*. 2008; **15**(9): 668–676.
51. Baeyens V. Clinical evaluation of bioadhesive ophthalmic drug inserts (BODI®) for the treatment of external ocular infections in dogs. *Journal of Controlled Release*. 2002; **85**(1–3): 163–168.

52. Batchelor H. Formulation strategies in mucosal-adhesive drug delivery. *CRS Newsletter*. 2005; **22**: 4–5.
53. Robinson JR, Bologna WJ. Vaginal and reproductive system treatments using a bioadhesive polymer. *Journal of Controlled Release*. 1994; **28**(1–3): 87–94.
54. O'Hagan DT, Rafferty D, Wharton S, Illum L. Intravaginal immunization in sheep using a bioadhesive microsphere antigen delivery system. *Vaccine*. 1993; **11**(6): 660–664.
55. Bonucci E, Ballanti P, Ramires PA, Richardson JL, Benedetti LM. Prevention of ovariectomy osteopenia in rats after vaginal administration of Hyaff 11 microspheres containing salmon calcitonin. *Calcified Tissue International*. 1995; **56**(4): 274–279.
56. Alam MA, Ahmad FJ, Khan ZI, Khar RK, Ali M. Development and evaluation of acid-buffering bioadhesive vaginal tablet for mixed vaginal infections. *AAPS PharmSciTech*. 2007; **8**(4): E109.
57. Cevher E, Taha MAM, Orlu M, Araman A. Evaluation of mechanical and mucoadhesive properties of clomiphene citrate gel formulations containing carbomers and their thiolated derivatives. *Drug Delivery*. 2008; (15): 57–67.
58. Available from: http://www.clinicaltrials.gov. [Last accessed on 2011 Nov 30].
59. Francois M, Snoeckx E, Putteman P, Wouters F, De Proost E, Delaet U, et al. A mucoadhesive, cyclodextrin-based vaginal cream formulation of itraconazole. *AAPS pharmSci*. 2003; **5**(1): E5.
60. Kim JK, Kim MS, Park JS, Kim CK. Thermo-reversible flurbiprofen liquid suppository with HP-beta-CD as a solubility enhancer: Improvement of rectal bioavailability. *Journal of Inclusion Phenomena and Macrocyclic Chemistry* 2009; **64**: 265–272.
61. Woolfson AD, McCafferty DF, McCarron PA, Price JH. A bioadhesive patch cervical drug delivery system for the administration of 5-fluorouracil to cervical tissue. *Journal of Controlled Release*. 1995; **35**(1): 49–58.
62. Woolfson A. Development and characterisation of a moisture-activated bioadhesive drug delivery system for percutaneous local anaesthesia. *International Journal of Pharmaceutics*. 1998; **169**(1): 83–94.
63. Donnelly RF, McCarron PA, Zawislak A, Woolfson AD. *Photodynamic Therapy of Vulval Neoplasias and Dysplasias: Design and Evaluation*

of Bioadhesive Photosensitiser Delivery Systems. VDM Verlag Dr Muller, Saarbrücken, Germany 2009. p. 1–400.

64. Schumacher U, Schumacher D. Functional histology of epithelia relevant for drug delivery. In *Bioadhesive Drug Delivery Systems* (Mathiowitz E, Chickering DE, III, Lehr C-M, eds) Marcel Dekker, Inc.: New York, 1999; 67–83.
65. Alam MA, Ahmad FJ, Khan ZI, Khar RK, Ali M. Development and evaluation of acid-buffering bioadhesive vaginal tablet for mixed vaginal infections. *AAPS PharmSciTech*. 2007; **8**(4): E109.
66. Salman HH, Gamazo C, Agüeros M, Irache JM. Bioadhesive capacity and immunoadjuvant properties of thiamine-coated nanoparticles. *Vaccine*. 2007; **25**(48): 8123–8132.
67. Junginger HE, Verhoef JC, Thanou M. Drug Delivery: Mucoadhesive Hydrogels. In *Encyclopedia of Pharmaceutical Technology*, 3rd ed., 2006. pp. 1169–1182.
68. Hassan EE, Gallo JM. A simple rheological method for the in vitro assessment of mucin-polymer bioadhesive bond strength. *Pharmaceutical Research*. 1990; **7**(5): 491–495.
69. Rossi S. Influence of mucin type on polymer-mucin rheological interactions. *Biomaterials*. 1995; **16**(14): 1073–1079.
70. Hägerström H, Paulsson M, Edsman K. Evaluation of mucoadhesion for two polyelectrolyte gels in simulated physiological conditions using a rheological method. *European Journal of Pharmaceutical Sciences*. 2000; **9**(3): 301–309.
71. Rossi S, Ferrari F, Bonferoni MC, Caramella C. Characterization of chitosan hydrochloride–mucin rheological interaction: Influence of polymer concentration and polymer: Mucin weight ratio. *European Journal of Pharmaceutical Sciences*. 2001; **12**(4): 479–485.
72. Hägerström H, Edsman K. Limitations of the rheological mucoadhesion method: The effect of the choice of conditions and the rheological synergism parameter. *European Journal of Pharmaceutical Sciences*. 2003; **18**(5): 349–357.
73. Chung HS, Park H, Kelly P, Robinson JR. Bioadhesive polymers as platforms for oral controlled drug delivery II: Synthesis and evaluation of some swelling, water-insoluble bioadhesive polymers. *Journal of Pharmaceutical Sciences*. 1985; **74**(4): 399–405.
74. Davis SS. The design and evaluation of controlled release systems for the gastrointestinal tract. *Journal of Controlled Release*. 1985 Nov [cited 2011 Nov 29]; **2**: 27–38.

75. Kremser C, Albrecht K, Greindl M, Wolf C, Debbage P, Bernkop-Schnürch A. In vivo determination of the time and location of mucoadhesive drug delivery systems disintegration in the gastrointestinal tract. *Magnetic Resonance Imaging*. 2008; **26**(5): 638–643.

76. Park K, Park H. Test methods of bioadhesion, in *Bioadhesive drug delivery systems*. (Lenaerts V, Gurney R, ed), Florida, Boca Raton, CRC Press; 1990.

77. McCarron PA, Donnelly RF, Zawislak A, Woolfson AD, Price JH, McClelland R. Evaluation of a water-soluble bioadhesive patch for photodynamic therapy of vulval lesions. *International Journal of Pharmaceutics*. 2005; **293**(1–2): 11–23.

78. McCarron PA, Donnelly RF, Zawislak A, Woolfson AD. Design and evaluation of a water-soluble bioadhesive patch formulation for cutaneous delivery of 5-aminolevulinic acid to superficial neoplastic lesions. *European Journal of Pharmaceutical Sciences* 2006; **27**(2–3): 268–279.

79. Donnelly RF, McCarron PA, Zawislak AA, Woolfson AD. Design and physicochemical characterisation of a bioadhesive patch for dose-controlled topical delivery of imiquimod. *International Journal of Pharmaceutics*. 2006; **307**(2): 318–325.

80. Salamat-Miller N, Chittchang M, Johnston TP. The use of mucoadhesive polymers in buccal drug delivery. *Advanced Drug Delivery Reviews*. 2005; **57**(11): 1666–1691.

81. McCrudden MTC, Thakur RRS, Migalska K, Donnelly RF. Strategies for enhanced peptide and protein delivery. *Therapeutic Delivery*. 2013; 4: 593–614.

82. Park, K. A new approach to study mucoadhesion: Colloidal gold staining. *International Journal of Pharmaceutics*. 1989; **53**(3): 209–217.

83. Davidovich-Pinhas M, Harari O, Bianco-Peled H. Evaluating the mucoadhesive properties of drug delivery systems based on hydrated thiolated alginate. *Journal of Controlled Release*. 2009; **136**(1): 38–44.

84. Davidovich-Pinhas M, Bianco-Peled H. Physical and structural characteristics of acrylated poly(ethylene glycol)–alginate conjugates. *Acta Biomaterialia*. 2011; **7**(7): 2817–2825.

85. Zambito, Y. Nanoparticles based on chitosan derivatives. In *Advances in Biomaterials Science and Biomedical Applications* (ed Pignatello R) InTech, New York, USA, 2013.

86. Gurny R, Meyer JM, Peppas NA. Bioadhesive intraoral release systems: Design, testing and analysis. *Biomaterials*. 1984; **5**, 336–340.

Chapter 11

Preventing Adhesion on Medical Devices

Pilar Teixeira and Fernanda Gomes

Institute for Biotechnology and Bioengineering,
Centre of Biological Engineering, University of Minho,
Campus de Gualtar, 4710-057 Braga, Portugal

pilar@deb.uminho.pt

The adhesion and biofilm formation ability of several important pathogens continues to be a problem for medical devices and one of the most common reasons of implant failures. Due to this fact, significant necessity exists of medical devices that can intrinsically resist to microbial colonization. In recent years, several approaches have been investigated and generally present an effective means to reduce microbial adhesion and thus prevent biomaterial-associated infections. These strategies involve the modification of surfaces mainly by the application of biomaterials coatings. This chapter focuses on several of these strategies.

11.1 Introduction

The utilization of implantable medical devices for short duration or intermittently, for months and years, such as an intravenous catheter or bone plates and rod, or permanently, as in an artificial

Bioadhesion and Biomimetics: From Nature to Applications
Edited by Havazelet Bianco-Peled and Maya Davidovich-Pinhas

ISBN 978-981-4463-98-0 (Hardcover), 978-981-4463-99-7 (eBook)
www.panstanford.com

heart valve or hip prosthesis, has increased significantly over the recent decades. These devices allow monitoring of vital functions and administration of drugs to critically ill patients, and intermittently or permanently support or replace failing organ functions. Thus, they help in saving human lives and greatly enhancing patients' quality of life. Materials for medical devices can be characterized as synthetic polymers, biodegradable polymers, bioactive materials, natural macromolecules (i.e., biopolymers), metals, carbons, and ceramics [45]. It is estimated that over 5 million medical devices or implants are used per year in the United States alone—approximately 350,000 pacemakers and 140,000 implantable cardioverter-defibrillator [9]. The most commonly used implants worldwide are artificial eye lenses; ear tubes; coronary stents; artificial knees; metal screws, pins, plates, and rods (traumatic fracture repair); intra-uterine devices; spine screws, rods, and artificial discs (spinal fusion hardware); breast implants; heart pacemakers and artificial hips. This consumption makes the medical devices companies across the world bring in over $200 billion in revenue per year. However, these devices are easily colonized with bacteria and fungi causing infection. The most frequent organisms isolated are Coagulase Negative Staphylococci (CoNS), particularly *S. epidermidis*, *S. aureus*, *Pseudomonas aeruginosa*, *Escherichia coli*, Streptococci, and Candida species [28]. Polymicrobial infections are also common. As some bacteria reside on the skin and mucous membranes, they are of low virulence potential to healthy individuals but strongly dangerous for patients with indwelling or implanted foreign polymer bodies and immunocompromised patients, such as premature babies and patients hospitalized for chemotherapy, other malignant diseases or organ transplantation [51]. Such infections cause suffering to patients, prolong the duration of hospital stays, increase antibiotic use and bacterial resistance, require additional diagnostic and therapeutic interventions, which generate added health care costs, and have other direct and indirect economic implications (loss of productivity and disability). According to the International Nosocomial Infection Control Consortium (INICC) report, which summarizes data of 422 intensive care units (ICUs) of 36 countries in Latin America, Asia, Africa, and Europe, for 2004–2009, the crude unadjusted excess mortalities of device-related infections ranged from 7.3% (for catheter-associated urinary tract infection) to 15.2% (for ventilator-associated pneumonia). The prevention of these infections is crucial, being the first step to avoid bacterial adhesion to the devices. The problem is aggravated when a biofilm is formed on the surface. Biofilms are complex consortia of adherent microorganisms encased in a polymeric matrix [42] that confers protection to the organisms against antimicrobial agents. In fact, biofilms are known to be between 100 to 1000 times more tolerant to antimicrobials than their planktonic counterparts, constituting thus an important bacterial survival strategy [12]. Besides, standard antibiotic therapy has been determined based on planktonic cells susceptibility, and therefore, in biofilm-related infections such therapies are usually not effective, allowing the propagation of sessile forms and their dissemination even after the end of the antibiotic administration. In recent years, considerable progress has been made in the prevention of biomedical devices associated infections and novel technologies have been introduced. This book chapter discusses the main strategies that have been employed to prevent bacterial adhesion to medical devices and thus medical device–related infections.

11.2 Adhesion and Biofilm Formation on Medical Devices

In natural, industrial, and medical environments, microorganisms are able to adhere to biotic and abiotic surfaces and grow in biofilm communities, being the predominant mode of growth of bacteria [52]. In fact, bacteria are able to attach to several kinds of surfaces, including metals, plastics, implant materials, and tissues [54]. Biofilm formation begins with the initial adhesion and subsequent agglomeration into multicellular structures. Hence, the adhesion of microorganisms to the surface of medical devices is considered the first step of implant infections and this basic pathogenic mechanism is the primary event of biofilm formation. The process of bacterial adhesion consists of two stages: an initial physicochemical interaction phase (phase one) and a next molecular and cellular interaction phase (phase two). In the first phase, bacteria move to or are moved to a material surface through physical forces, such as Brownian motion, van der Waals attraction forces, gravitational forces, effect of surface electrostatic charge, and hydrophobic interactions or by chemotaxis. These physical interactions are classified as long-range interactions (nonspecific, distances >50 nm), which are a function of the distance and free energy, and short -range interactions (<5 nm). Bacteria are transported to the surface by the first interactions while upon closer contact, short-range interactions become more important. The latter can be separated into chemical bonds (such as hydrogen bonding), ionic and dipole interactions and hydrophobic interactions. The second phase of adhesion is governed by molecular-specific reactions between bacterial surface structures and substratum surfaces. In this phase, bacteria strongly adhere to a surface through bacterial surface polymeric structures, such as fimbriae, capsules, or pili and slime. Furthermore, this phase is also affected by other factors like the physicochemical properties of both microbial cells and solid surfaces, and environmental factors such as the flow conditions and the presence of antibactericides. The most relevant surface properties are hydrophobicity, surface charge, material surface topography/ roughness, and surface chemical composition. Concerning the effect of hydrophobicity, bacteria with hydrophobic properties adhere in a higher extent to hydrophobic materials, while hydrophilic bacteria prefer hydrophilic surfaces. Besides, hydrophobic bacteria adheres in a higher number than hydrophilic bacteria. Most particles acquire an electric charge in aqueous medium due to the ionization of their surface groups. In this way, as bacteria and the surface materials in aqueous suspension are almost negatively charged, usually adhesion is not ruled by electrostatic interactions. Surface roughness affects bacterial adhesion because a rough surface has a greater surface for bacteria adhere and the depressions allow bacteria to become defended from shear stress. However, it must be noted that if the surface presents a microroughness, the minimal number of contact points may reduce the possibility of adhesion. If the surface chemical composition is modified, for instance, by surface coatings, bacterial adhesion is avoided.

Disruptive forces are crucial for the formation of fluid-filled channels, whose function is oxygen, metabolites and nutrient transport to all cells. Moreover, disruptive forces are also involved in the detachment of cells promoting the dissemination of infection [66]. This step of biofilm formation can promote the colonization of new surfaces in other infection sites and then lead to the systemic propagation of infection. The ability of several microorganisms to adhere and consequently form biofilm is a crucial virulence factor that allows them to resist host immune system and antibiotics. Cells in this lifestyle are nowadays recognized as having an increased tolerance to antimicrobial agents and to innate host defense, such as to antimicrobial peptides (AMPs) and neutrophil phagocytosis [20].

11.3 Strategies to Prevent Adhesion on Medical Devices

The prevention, consisting of sterilization of medical devices and equipment as well as body parts of patients and health care personnel, and other prophylactic measures, is of great importance and the best way to deal with medical device infections. However, these measures are often insufficient to prevent this kind of infections. The adhesion of cells to medical surfaces is a preponderant process to biofilm formation and consequently to medical device infections. In accordance with this notion, many researchers work hard aiming to develop the most effective approaches able to prevent the colonization of medical devices. For that, several strategies focusing on surfaces treatment are being investigated. These strategies generally consist of the modification of the surface with hydrophilic non-fouling polymeric surfaces coatings, minimizing the protein adsorption and microbial adhesion [53]. Therefore, the surface modification and the incorporation of antimicrobial agents in the bulk material or as a surface coating, such as antibiotics, peptides, quorum-sensing molecules, natural compounds, nanoparticles, and bacteriophages are examples of some approaches studied in terms of their impact on bacterial adhesion and accumulation. These strategies having the prevention of adhesion as target work as prophylactic measures, aims to avoid/decrease the incidence of medical device related infections complementing the hygiene and sterilization care practice.

11.3.1 Current Recommendations

The main responsibility for infection of medical implants and devices is attributed to hospitals, namely to the medical staff and also to patients. In fact, these infections arise from the hands of the surgical staff during implantation/application; a contaminated implant/device surface; the patient's own skin or mucus membrane; a distant local infection in the patient; contaminated disinfectants; and contact with other patients in the hospital or family members after intervention [6,72]. These risks can be overcome or at least minimized by the practice of stricter hygiene procedures in hospitals. For that, it is vital to follow the specific guidelines to each type of catheter as the Guidelines for the Prevention of Intravascular Catheter-Related Infections [65] or the Guidelines for Prevention of Catheter-Associated Urinary Tract Infection [39], both developed by Center for Disease Control and Prevention (CDC) and Healthcare Infection Control Practices Advisory Committee (HICPAC). These guidelines were developed for healthcare personnel who insert intravascular catheters and for persons responsible for the surveillance and control of infections in hospital, outpatient, and home healthcare settings. However, infections associated with medical devices can never be completely avoided and thus strategies to avoid infection by inhibiting bacterial adhesion have to be followed.

11.3.2 Emergent Approaches

Medical device–associated infections still occur in spite of careful hygienic procedures in their application and post application. In recent years, several studies have been focused on the development of antibacterial surfaces in order to control bacterial adhesion and biofilm formation. Some of these promising strategies are described below.

11.3.2.1 Surface topography modification

Catheter surface modification is one of the approaches studied to prevent cell adhesion and biofilm formation. This modification aims to alter the characteristics of the surfaces in order to obtain a non-biofilm forming area and consists of changes of size and porous distribution and roughness. Bacterial adhesion requires an initial physicochemical interaction phase and is a process that depends on many factors such as bacterial properties, the surface characteristic, namely surface topography, and environmental factors [51]. Scheuerman et al. (1998) observed that irregular polymeric surfaces were more prone to bacterial adhesion and biofilm deposition than smooth surfaces [76]. A possible explanation is that roughened surfaces present an increased surface area and then provide more favorable sites for colonization. Díaz et al. (2007) showed that the substratum nature has an important influence in bacterial adhesion [27]. Bacterial adhesion was more significant on smooth surfaces than microrough ones. On the other hand, nanopatterned surfaces were more adhered than smooth ones. These authors also observed that bacterial adhesion was dependent of the relation between roughness characteristics dimensions and bacterial size. Comparing smooth and nanostructured surfaces significant differences were observed in morphology, orientation, and distribution of bacteria [27].

Mitick-Dineva et al. (2009) tested the different ability of *E. coli*, *P. aeruginosa*, and *S. aureus* to adhere to glass surfaces modified [60]. This modification consisted of changing the surface topography at the nanoscale, reducing the overall surface roughness parameters. All pathogens studied exhibited an increased propensity for adhesion to the nano-smooth surfaces.

11.3.2.2 Surface coatings

A surface coating can be defined as the buildup of an antibacterial material onto the substratum surface [36] and is one of the most used strategies to produce antibacterial surfaces. Antibacterial coatings act by either being toxic when coming into contact with the bacteria or by releasing an effective chemical or antibacterial agent from the surface [48,91].

The most common strategies for protein- or microorganisms-repellent coatings are polymer layers and brushes on surfaces [53]. Polymeric surfaces are mostly modified with brush layers via direct covalent coupling to reactive groups on the surface. These reactive groups can be designed into the original surface or generated by for instance plasma treatment of the surface. The higher the surface density of the polymer chains on the surface, the better the anti-adhesive characteristics will be. Also, the chain length of the polymers is of importance; longer chains are more effective in prevention of bacterial adhesion [44,71]. Because most of the polymers used for brushes are hydrophilic, water will be attracted into the brush layer and form a repellent layer close to the surface. The water is held in place via hydrogen bonding to the polymers. Consequently, there is steric hindrance for proteins or microorganisms to adsorb to the surface. The result is a reduction of protein adsorption by several orders of magnitude [53].

Quaternary ammonium compounds (QACs) are also widely applied as antibacterial agents [61]. The main antimicrobial activity of QACs is due to their surfactant characteristic, which provokes the distortion of the membranes of the microorganisms, leakage and cell death. In spite of QAC coatings possess a long-lasting contact-based antibacterial mechanism, it has been reported that certain bacteria are able to develop resistance against these surfaces [88].

Silver-based coatings are also widely used in medical implants due to the silver ions released from the surface being bactericidal against both Gram-positive and Gram-negative bacteria [48,53].

Likewise, hydroxyapatite (HA) coatings provide medical devices surfaces such as titanium with antibacterial properties. Unfortunately, HA coatings have the drawback that they are mechanically weak and can be nonuniform in density and thickness [81].

Other possible surface modification is to coat the device with hydrophilic negatively charged polysaccharides such as hyaluronan [67] and heparin [62]. In a work aiming to study the adhesion ability of *S. epidermidis* to acrylic and silicone surfaces preconditioning with heparin and gentian violet it was observed a significant reduction (53–90%, $P < 0.05$) in bacterial adhesion on the preconditioning surfaces [86].

Altering the surface chemistry is other approach to prevent bacterial adhesion. Surface of devices with different functional groups change bacterial adhesion since it depends on material hydrophobicity, charge, and roughness. Several studies on self-assembled monolayers (SAMs) to control interfacial characteristics have been reported [35,80,89,94]. Sousa and co-authors (2008) assessed the effect of gold coating on staphylococcal adhesion, as well as of subsequent coverage with SAMs with a methyl terminal group and hydrophobic character and hydrophilic SAMs with a carboxylic acid terminal group [86]. These authors observed the lowest levels of *S. epidermidis* adhesion in addition to gold covered surfaces, on the methyl terminated SAMs, octanethiol (OT) and hexadecanethiol (HDT). However, this technology has relatively poor chemical stability and an impossible control over the quality of the monolayer [79].

The use of surface coatings as antibacterial agents has still revealed several limitations. Bacteria can develop resistance to these agents and the durability of the target surface may not be sufficient to maintain long-term antibacterial behavior [43].

11.3.2.3 Drug-delivery coatings

An emerging new trend to reduce bacterial adhesion is based on coatings that actively release antimicrobial agents (antibacterial, disinfectant, or metal ions). Such "active" coatings have been designed to release an effective dose locally for an extended period, via a "slow-delivery system." At the implant surface itself, the drug diffusing out of the device becomes highly concentrated within the so-called stagnant diffusion layer (Nernst layer) exceeding usual systemic antibiotic concentrations by several orders of magnitude, thereby avoiding systemic effects [77]. During the critical short-term post-implantation period (several hours) high initial fluxes of antibacterial agents are necessary to inhibit initial bacterial adhesion, while a continued release beyond this short-term period is desirable because protective fibrous capsule formation and tissue integration occur over a longer time period (weeks to months) [46]. Metals, especially copper and silver ions [7,22]; disinfectants such as chlorhexidine [98] and other bactericidal chemicals such as nitric oxide [46]; and various antibiotics, including gentamicin, vancomycin, and others [22,30,49,63,96,99] are the most released antimicrobial agents from coatings. Among metals, silver is the most used due to its low toxicity and good antimicrobial action [59,83].

The local delivery of antibiotics can prevent bacterial adhesion and posterior growth of a significant number of bacteria and, consequently, of biofilm formation. This strategy has been long applied in bone cements that are used to fix orthopedic and orthodontic implants [49,96]. Usually, an antibiotic like gentamycin is mixed into the cement and subsequently slowly released after hardening in situ. Antibiotics can also be contained on biodegradable surface coatings [63,99]. In this case, the surface layer would slowly degrade and cause antibiotic release directly at the surface-tissue interface. The kinetics of antibiotic release closely follows the degradation kinetics of the coating. This approach has been tested for orthopedic applications.

Coatings containing combinations of antibiotics and antiseptics have also been used in the internal and external surface of catheters, e.g., minocycline and rifampin or chlorohexidin and silver-sulfadiazine [22,30].

Some disadvantages of this approach are the use of active agents, which may be associated with side effects such as cytotoxicity, hypersensibility, inflammatory responses, or the progressive alarming phenomenon of antibiotic resistance [37].

11.3.2.4 Antimicrobial peptides

Antimicrobial peptides are emerging as potential antimicrobial agents in the control of medical device infections due to their broad spectrum of activity, to the absence of resistance development and to their rapid mechanism of action. They are generally short molecules, positively charged and present ≥30% of hydrophobic residues. Due to their amphiphilic properties and helical structures and their ability to form pores in microbial membranes, they are able to promote membrane rupture and cell death. Furthermore, nowadays it is known that AMPs can have an important role in the inhibition of the biosynthesis of proteins, nucleic acids, microbial cell walls and are also able to inhibit the activity of some microbial enzymes [73]. These antimicrobial agents can be applied as single or multilayer coatings. Generally, each layer of antimicrobial peptides is followed by a biodegradable polymer layer, which is progressively degraded by the action of the human environment releasing gradually the antimicrobial peptide. This application of antimicrobial peptides has several advantages, namely high biocompatibility, low toxicity, and well-controlled release [50]. Moreover, unlike several antibiotics, AMPs do not depend on the presence of active cells to exercise their effect. They are able to affect inactive or slow-growth bacteria [5]. These molecules are capable of permeabilizing and forming pores within cytoplasmic membranes modifying their permeability and stability. Lantibiotics, temporins, cathelicidins, and defensins are some examples of the most explored AMPs for clinical use.

One of the approaches studied aiming to improve the inhibitory bacterial adhesion of AMPs was the covalent immobilization of these molecules, which promotes the reduction of the microorganism survival after contact with the coated material [18].

Magainin, an antimicrobial peptide, photochemically immobilized in a silicone rubber surface was able to inhibit by 98% *S. epidermidis* growth [82].

In another study, Cirioni et al. (2006) demonstrated that cathelicidin peptide BMAP-28 was effective against *S. aureus* [16]. Moreover, they showed good attachment efficiency to biomaterial surfaces. Both these features make this AMP a potential coating to be used into the control of medical device–associated infections promoted by Staphylococci.

Melimine, a synthetic peptide used as coating of contact lenses, reduced the adhesion of *P. aeruginosa* and *S. aureus* [97]. Tsai et al. (2011) suggested the use of LL-37, a human cathelicidin, as therapeutic or preventive option against *C. albicans* infections since this molecule prevented the adhesion of this fungus to polystyrene and tissues [93]. This was suggested to be due to its interaction with *C. albicans* surface namely with cell-wall carbohydrates, promoting cell aggregation and consequently their incapacity to adhere to the supporting substrata [73].

Lysostaphin, a bacterial antimicrobial peptide, showed to inhibit *S. aureus* adhesion to hernia repair meshes evidencing a promising alternative in the prevention of prosthesis infections [75].

In general, AMPs demonstrated to possess several advantages as biomedical surfaces coatings and seems to be suitable for the design of non-biofilm surfaces.

11.3.2.5 Quorum-sensing molecules

Another approach is the use of molecules that interfere with Quorum-sensing (QS). Quorum-sensing is a complex regulatory process that involves cell–cell communication via the secretion of chemical signals called autoinducers [33]. Some examples of QS molecules are oligopeptides and *N*-acyl homoserine lactones in Gram-positive and Gram-negative, respectively, and a family of signal molecules called Al-2 in both types of bacteria. QS process has an important role in modulating the expression of genes related with the production of specific enzymes, virulence factors and metabolites as well as in the formation of microbial communities as biofilms. By regulating the expression of specific genes this cell-to-cell communication allows a response to the needs presented by sessile microorganisms namely those associated to the increasing population density [33].

Dell'Acqua et al. (2004) were the first that directly demonstrated that the interference of cell–cell communication by QS inhibitors can eliminate medical device-related infections namely those promoted by staphylococci [25]. They tested RNAIII-inhibiting peptide (RIP) against *S. aureus* and *S. epidermidis* verifying that this molecule is able to completely inhibit adhesion and thus prevent biofilm formation. The inhibition of the phosphorylation of RIP's target proteins promotes the disruption of cell-to-cell communication and thus QS process [25].

RIP was then suggested as a possible alternative to antibiotics as coating of medical devices in order to prevent infections. In addition to report the effect of RIP-coated central venous catheters in the reduction of bacterial adhesion of *S. aureus*, Cirioni et al. (2006) demonstrated the synergistic effect of RIP in combination with antibiotics against *S. aureus* [16]. When tested in orthopedic implants and ureteral stents, RIP inhibited the in vivo adhesion and biofilm formation of *S. aureus* [2,17].

Furanones are another example of this kind of molecules. Lönn-Stensrud et al. (2009) demonstrated that this QS inhibitor prevented biofilm formation of *S. epidermidis* [57], being also able to control multicellular behavior in Gram-negative microorganisms [58,69]. These molecules are usually used in materials employed in implantable devices namely silicone, expanded polytetrafluoroethylene and polypropylene [84].

11.3.2.6 Nanoparticles

Modification of medical device surfaces with nanoparticles (NPs) was recently studied in the prevention of bacterial adhesion. Nanoparticles are nowadays regarded as promising antimicrobial agents due to the high surface area to volume ratio and their unique chemical and physical characteristics [68]. Although other NPs have already been studied (copper, gold, magnesium, etc.), silver NPs have been the most effective, presenting a good antimicrobial activity against several microorganisms [38]. There are some studies reporting the efficiency of these NPs in the prevention and reduction of biofilm formation on catheter devices [47,70,74]. However, the exact mechanism of action of this agent is not yet known. Another problem is the great variation of the efficiency of silver on medical devices [53]. Selenium NPs are another strategy recently developed as coating of medical devices. Tran and Webster [92] obtained good results using this coating in common polymeric substrates (polyvinyl chloride (PVC), polyurethane and silicone).

Selenium NPs were able to inhibit bacterial growth of *S. aureus*. In this study, they also compared the efficiency of selenium and silver NPs on PVC substrates and verified that selenium was the most effective.

Lellouche and coworkers (2012) showed in their study that yttrium fluoride (YF3) nanoparticles exhibited antibacterial activity against *E. coli* and *S. aureus* and this was dependent on NP size having the smaller size a higher antibacterial effect than larger NPs [56]. The modification of catheter surfaces with these NPs prevented the adhesion and consequent colonization by these two common bacterial pathogens. Once again, these results highlight the promising use of these metal fluoride NPs as novel antimicrobial and antibiofilm agents and as medical devices coatings taking advantage of their low solubility and providing extended protection [56].

Chifiriuc et al. (2012) tested *Rosmarinus officinalis* essential oil coated NPs pelliculised on the surface of catheter pieces [14]. Using this strategy, these authors studied the combined action of NPs and a natural compound known for its antimicrobial properties. This nanobiosystem exhibited high efficiency in inhibiting *C. albicans* and *C. tropicalis* adhesion and biofilm accumulation. These results reinforced the potentialities of the use of nanoparticles in the medical field. Furthermore, the combination of anti-adhesion strategies can potentiate an improvement in the resistance to microbial colonization and biofilm formation opening new directions for the design of film-coated surfaces.

11.3.2.7 Bacteriophages

For a long time, phage therapy has attracted attention of several groups of investigation, but nowadays there is a renewed interest in this therapy namely in the control of adhesion and biofilm formation [29]. Bacteriophages are viruses that infect, replicate within the host, and kill bacteria [13]. Therefore, in the first stage called "penetration," the virus injects its genome into the host and begins to replicate and multiply inside the bacterial cell. At the end of their life cycle, phages promote bacterial lyse releasing newly formed phage particles that are free to infect other cells [73]. Kutateladze and Adamia (2010) demonstrated that these viruses are anti-infectives in humans and animals [55] and Fischetti (2010) also demonstrated that some lytic phages can be used as antimicrobials against pathogenic bacteria [32]. Since they are harmless to humans [8], they gained a significant importance especially in clinical use. Their emergence arose from the necessity of new alternatives to antibiotics in order to overcome the problem of resistance. One example of phages studied as potential alternative to antibiotics and presenting anti-biofilm properties are phages specific to *S. epidermidis*. These phages were able to destroy biofilm by extracellular material degradation. In fact, some phages produce depolymerases, whose target is the exopolysacharide (EPS), which hydrolyses and degrades the extracellular polymeric matrix of biofilms [29,41].

The use of phages has several advantages such as the strong antibacterial activity, the absence of adverse effect after phage therapy [4,40,87], and rapid, simple, and inexpensive phage production [4]. However, this approach has also some disadvantages, namely the narrow lytic spectrum and the low probability of finding a specific phage with high lytic action and expressing relevant exopolymeric degrading enzymes [4]. To circumvent this problem, a phage cocktail can be a solution, consisting of the combination of several phages aiming to create an optimal mixture with broad activity against host cells. Another alternative are engineered phages that are able to degrade medical biofilms and alter their host range in order to make them able to infect a wide range of host bacterial strains and species. Moreover, their manipulation also consists of the introduction of new genes aiming the induction of the production of specific enzymes essential for the degradation of the biofilm matrix, facilitating their entrance into the biofilm. Taking into consideration their capacity to avoid biofilm accumulation on medical devices, their possible use as therapeutic strategy gained importance mainly in the area of catheter coatings [73].

Some researchers investigated the in vitro use of phages by incorporating these microorganisms into a hydrogel used to coat medical devices. Curtin and Donlan (2006) treated silicone catheter using phage 456 with and without supplemental divalent cations [21]. They obtained a reduction in *S. epidermidis* biofilm viable cells of 4.47 and 2.34 log CFU/cm^2, respectively. On the other hand, Fu et al. (2010) tested a hydrogel impregnated with a cocktail of phages in Foley catheters and studied their effect in adhesion and biofilm formation of *P. aeruginosa* [34]. They observed a significant mitigation of biofilm formation by these clinical relevant bacteria. Overall, this pre-treatment coating was able to attenuate adhesion and reduce the biofilm formation of *S. epidermidis* and *P. aeruginosa* [21,34].

Carson et al. (2010) also studied the effect of phage-coating onto Foley catheter biomaterials on *E. coli* and *Proteus mirabilis* adhesion and biofilm formation [10]. Once again, this pre-treatment was able to prevent biofilm formation and reduce the number of viable cells on the surface of *circa* of 90% in both *E. coli* and *P. mirabilis*. On the other hand, Fenton et al. (2013) showed the efficacy of CHAPK, a bacteriophage-derived peptidase, as biocidal agent for prevention and treatment of biofilm-associated staphylococcal infections namely *S. epidermidis* and all known clonal types of Methicillin-resistant *S. aureus* [31]. Therefore, they suggested their utilization as a coating agent on medical implants for prevention of infections caused by the aforementioned species. Thus, phages can be used as a prophylactic measure in the pre-treatment of medical devices and decrease the incidence of catheter-associated infections [31,84].

11.3.2.8 Natural compounds

Natural compounds used as coating surfaces of medical devices and presenting anti-adherence characteristics are urgently needed in the medical field. In recent years, several researchers suggested the possible use of broad-range natural antimicrobial agents in clinical setting as a potential alternative to conventional antibiotics. For example, heparin, a highly sulfated glycosaminoglycan, immobilized on plastic surfaces was shown to reduce bacterial adherence in vitro and in vivo [3,90]. On the other hand, essential oils of Boswellia spp., effectively inhibit the development of biofilms of medically relevant species, namely *S. aureus*, *S. epidermidis*, and *C. albicans* [78]. Another essential oil was studied by Cavalcanti et al. (2011), who demonstrated that the essential oil of *R. officinalis* (Rosemary) has anti-adherent activity against *C. albicans* [11]. Further, a study involving *M. piperita* essential oil reported the anti-adhesive properties of this natural substance when used as coating of prosthetic devices [1]. In addition to present anti-adherence properties against *S. aureus*, a microbicidal effect was also observed. This effect results from the stabilization, decrease of volatility, and controlled release of the essential oil from the core/shell nanostructure. Taking into account these results, these authors suggested this kind of strategies as good candidates for the design of novel material surfaces used for prosthetic devices.

Cinnamaldehyde and carvacrol also exhibited a significant antimicrobial activity and consequently an anti-biofilm effect against *E. coli*, *P. aeruginosa* and *S. aureus*. When incorporated into poly(lactic-co-glycolic acid) polymer films, these natural antimicrobial compounds were able to mitigate biofilm formation of *S. aureus* and *E. coli*. In conclusion, they proposed the use of these two natural compounds as plant-based coating for indwelling devices to safely prevent adhesion and consequent accumulation by pathogenic bacteria [100].

Nowatzki et al. (2012) developed a new strategy to impede bacterial adhesion and biofilm growth on synthetic materials [64]. This approach is based in the controlled salicylic acid release from a polymeric coating. This kind of coating was effectively inhibitory against *E. coli* and *P. aeruginosa*. When mimicking the physiological urine flow, urinary catheters coated with the salicylic acid-polymer exhibited a significant reduction of biofilm formation by *E. coli*.

11.3.2.9 Electrical methods

Electrical approaches have been used either to release antimicrobials from device surfaces and thus prevent biofilm formation or to drive antimicrobials through established biofilms. This last method, the bioelectric effect, consists of the concurrent application of antibiotics and a weak electric field to kill bacteria. According to Costerton and co-authors (1994), with the application of direct current electric fields between 1.5 and 20 V/cm, the concentrations of antibiotics needed to be effective against biofilm bacteria fell from approximately 5000 times to 4 times greater than those necessary for planktonic bacteria in the absence of electricity [19]. However, in a work that aimed to study the in vitro bioelectric effect on the activities of 11 antimicrobial agents representing a variety of different classes against *P. aeruginosa*, MRSA, and *S. epidermidis*, it was observed that the enhancement of the activity of antimicrobial agents against biofilm organisms by electrical current is not a generalizable phenomenon across microorganisms and antimicrobial agents [24]. It has been suggested that the mechanism of antibacterial activity of electrical current results from the oxidation of enzymes and coenzymes, membrane damage leading to the leakage of essential cytoplasmic constituents, toxic substances (e.g., H_2O_2, oxidizing radicals, and chlorine molecules) produced as a result of electrolysis, and/or a decreased bacterial respiratory rate [24]. It must be noted that this method is more a therapeutic one rather than a preventive measure.

11.4 Conclusions

A serious health problem is represented by the high occurrence of medical device infections, the colonization of these surfaces by microorganisms and biofilm formation being the major cause of this kind of infections. This issue was aggravated by the emergence of antibiotic resistance that prompted the search of novel agents able to inhibit bacterial adhesion and biofilm growth. In recent years, further efforts have been made by several groups of researchers aiming to develop new approaches for the design of sterile surface coatings that may be useful for various medical applications. Biomaterial surface properties including hydrophobicity and roughness must be taken into consideration since they control adhesion and growth of infecting bacteria. Here we mentioned several of these novel anti-biofilm strategies and several of them demonstrated promising applicability in medical environment since they exhibited great ability to counteract microbial adhesion and biofilm formation on medical devices surfaces. However, researchers are always looking for new and most effective strategies aiming the successful reduction or the total eradication of the infections focused here. To achieve this, more studies will be necessary, preferentially in vivo, to better understand the mechanisms of bacterial adhesion and biomaterials infection.

Acknowledgment

P. Teixeira and F. Gomes acknowledge the financial support of the Portuguese Foundation for Science and Technology through the grants SFRH/BPD/86732/2012 and SFRH/BPD/84488/2012, respectively.

References

1. Anghel, I., and Grumezescu, A. M. (2013). Hybrid nanostructured coating for increased resistance of prosthetic devices to staphylococcal colonization, *Nanoscale Res. Lett.*, **8**, 6.
2. Anguita-Alonso, P., Giacometti, A., Cirioni, O., Ghiselli, R., Orlndo, F., Saba, V., Scalise, G., Sevo, M., Tuzova, M., Patel, R., and Balaban, N. (2007). RNAIII-Inhibiting-peptide-loaded polymethylmethacrylate prevents in vivo *Staphylococcus aureus* biofilm formation. *Antimicrob. Agents Chemother.*, **51**, 2594–2596.
3. Appelgren, P., Ransjo, U., Bindslev, L., Espersen, F., and Larm, O. (1996). Surface heparinization of central venous catheters reduces microbial colonization in vitro and in vivo: Results from a prospective, randomized trial, *Crit. Care Med.*, **24**, 1482–1489.
4. Azeredo, J., and Sutherland, I. W. (2008). The use of phages for the removal of infectious biofilms, *Curr. Pharm. Biotechnol.*, **9**, 261–266.
5. Batoni, G., Maisetta, G., Brancatisano, F. L., Esin, S., and Campa, M. (2011). Use of antimicrobial peptides against microbial biofilms: Advantages and limits, *Curr. Med. Chem.*, **18**, 256–279.
6. Bearman, G. M. L., Munro, C., Sessler, C. N., and Wenzel, R. P. (2006). Infection control and the prevention of nosocomical infections in the intensive care unit, *Sem. Resp. Crit. Care Med.*, **27**, 310–324.
7. Borkow G., and Gabbay, J. (2004). Putting copper into action: copperimpregnated products with potent biocidal activities, *FASEB J.*, **18**, 1728–1730.
8. Bruttin, A., and Brussow, H. (2005). Human volunteers receiving *Escherichia coli* phage T4 orally: a safety test of phage therapy. *Antimicrob. Agents Chemother.*, **49**, 2874–2878.
9. Bryers, J. D. (2008). Medical biofilms, *Biotechnol. Bioeng.*, **100**, 1–18.
10. Carson, L., Gorman, S. P., and Gilmore, B. F. (2010). The use of lytic bacteriophages in the prevention and eradication of biofilms of *Proteus mirabilis* and *Escherichia coli, FEMS Immunol. Med. Microbiol.*, **59**, 447–455.
11. Cavalcanti, Y. W., Almeida, L. F. D., and Padilha, W. W. N. (2011). Anti-adherent activity of Rosmarinus officinalis essential oil on *Candida albicans*: An SEM analysis, *Rev. Odonto. Cienc.*, **26**, 139–144.
12. Ceri, H., Olson, M., Stremick, C., Read, R. R., Morck, D., and Buret, A. (1999). The Calgary biofilm device: New technology for rapid determination of antibiotic susceptibilities of bacterial biofilms, *J. Clin. Microbiol.*, **37**, 1771–1776.

13. Chibani-Chennoufi, S., Dillmann, M. L., Marvin-Guy, L., Rami-Shojaei, S., and Brussow, H. (2004). *Lactobacillus plantarum* bacteriophage LP65: A new member of the SPO1-like genus of the family Myoviridae, *J. Bacteriol.*, **186**, 7069–7083.

14. Chifiriuc, C., Vgrumezescu, V., Grumezescu, A. M., Saviuc, C., Lazăr, V., and Andronescu, E. (2012). Hybrid magnetite nanoparticles/ *Rosmarius officinalis* essential oil nanobiosystem with antibiofilm activity, *Nanoscale Res. Lett.*, **7**, 209.

15. Cirioni, O., Ghiselli, R., Minardi, D., Orlando, F., Mocchegiani, F., Silvestri, C., Muzzonigro, G., Saba, V., Scalise, G., Balaban, N., and Giacometti, A. 2007. RNAIII-inhibiting peptide affects biofilm formation in a rat model of staphylococcal ureteral stent infection, *Antimicrob. Agents Chemother.*, **51**, 4518–4520.

16. Cirioni, O., Giacometti, A., Ghisell,i R., Bergnach, C., Orlando, F., Mocchegiani, F., Slivestri, C., Licci, A., Skerlavaj, B., Zanetti, M., Saba, V., and Scalise, G. (2006). Pre-treatment of central venous catheters with the cathelicidin BMAP-28 enhances the efficacy of antistaphylococcal agents in the treatment of experimental catheter-related infection, *Peptides*, **27**, 2104–2110.

17. Cirioni, O., Giacometti, A., Ghiselli, R., Dell'Acqua, G., Orlando, F., Mocchegiani, F., Silvestri, C., Licci, A., Saba, V., Scalise, G., and Balaban, N. (200). RNAIII-inhibiting peptide significantly reduces bacterial load and enhances the effect of antibiotics in the treatment of central venous catheter-associated *Staphylococcus aureus* infections, *J. Infect. Dis.*, **193**, 180–186.

18. Costa, F., Carvalho, I. F., Montelaro, R. C., Gomes, P., and Martins, M. C. L. (2011). Covalent immobilization of antimicrobial peptides (AMPs) onto biomaterial surfaces, *Acta Biomater.*, **7**, 1431–1440.

19. Costerton, J. W., Ellis, B., Lam, K., Johnson, F., and Khoury, A. E., (1994). Mechanism of electrical enhancement of efficacy of antibiotics in killing biofilm bacteria, *Antimicrob. Agents Chemother.*, **38**, 2803–2809.

20. Costerton, J. W., Stewar,t P. S., and Greenberg, E. P. (1999). Bacterial biofilms: a common cause of persistent infections, *Science*, **284**, 1318–1322.

21. Curtin, J. J., and Donlan, R. M. (2006). Using bacteriophages to reduce formation of catheter-associated biofilms by *Staphylococcus epidermidis*, *Antimicrob. Agents Chemother.*, **50**, 1268–1275.

22. Darouiche, R. O. (1999). Anti-infective efficacy of silver-coated medical prostheses, *Clin. Infect. Dis.*, **29**, 1371–1377.

23. Darouiche, R. O., Raad, I. I., Heard, S. O., Thornby, J. I., Wenker, O. C., Gabrielli, A., Berg, J., Khardori, N., Hanna, H., Hachem, R., Harris, R. L., and Mayhall, G. (1999). A comparison of two anti-microbial-impregnated central venous catheters, *New Eng. J. Med.*, **340**, 1–8.
24. Del Pozo, J. L., Rouse, M. S., Mandreka, R J. N., Sampedro, M. F., Steckelberg, J. M., and Patel, R. (2009). Effect of electrical current on the activities of antimicrobial agents against Pseudomonas aeruginosa, *Staphylococcus aureus*, and *Staphylococcus epidermidis* biofilms, *Antimicrob. Agents Chemother.*, **53**, 35–40.
25. Dell' Acqua, G., Giacometti, A., Cirioni, O., Ghiselli, R., Saba, V., Scalise, G., Gov, Y., and Balaban, N. (2004). Suppression of drug-resistant staphylococcal infections by the quorum-sensing inhibitor RNAIII-inhibiting peptide, *J. Infect. Dis.*, **190**, 318–320.
26. Dhruva, S. S., Bero, L. A., and Redberg, R. F. (2009). Strength of study evidence examined by the FDA in premarket approval of cardiovascular devices, *JAMA*, **302**, 2679–2685.
27. Díaz, C., Cortizo, M. C., Schilardi, P. L., Saraiva, S. G. G., and Mele, M. A. F. L. (2007). Influence of the nano-micro structure of the surface on bacterial adhesion, *Mat. Res.*, **10**, 11–14.
28. Diekema, D. J., Pfaller, M. A., and Schmitz F. J. (2001). Survey of infections due to staphylococcus species: Frequency of occurrence and antimicrobial susceptibility of isolates collected in the United States, Canada, Latin America, Europe and the Pacific region for the Sentry antimicrobial surveillance program, 1997–1999, *Clin. Infect. Dis.*, **32**, 114–132.
29. Donlan, R. M. (2009). Preventing biofilms of clinically relevant organisms using bacteriophage, *Trends Microbiol.*, **17**, 66–72.
30. Falagas, M. E., Fragoulis, K., Bliziotis, I. A., Chatzinikolaou, I. (2007). Rifampicin-impregnated central venous catheters: A meta-analysis of randomized controlled trials, *J. Antimicrob. Chemother.*, **59**, 359–369.
31. Fenton, M., Keary, R., McAuliffe, O., Ross, R. P., O'Mahony, J., and Coffey, A. (2013). Bacteriophage-derived peptidase CHAPK eliminates and prevents staphylococcal biofilms, *Int. J. Microbiol.*, (doi: 10.1155/2013/625341).
32. Fischetti, V. A. (2010). Bacteriophage endolysins: A novel anti-infective to control Gram-positive pathogens, *Int. J. Med. Microbiol.*, **300**, 357–362.
33. Francolini, I., and Donelli, G. (2010). Prevention and control of biofilm-based medical-device-related infections, *FEMS Immunol. Med. Microbiol.*, **59**, 227–238.

34. Fu, W., Forster, T., Mayer, O., Curtin, J. J., Lehman, S. M., and Donlan, R. M. (2010). Bacteriophage cocktail for the prevention of biofilm formation by Pseudomonas aeruginosa on catheters in an in vitro model system, *Antimicrob. Agents Chemother.*, **54**, 397–404.

35. Fujimoto, H., Yoshizako, S., Kato, K., and Iwata, H. (2006). Fabrication of cell-based arrays using micropatterned alkanethiol monolayers for the parallel silencing of specific genes by small interfering RNA, *Bioconjug. Chem.*, **17**, 1404–1410.

36. Geesink, R., de Groot, K., and Klein, C. P. (1988). Bonding of bone to apatite-coated implants, *J. Bone Joint Surg. Br.*, **70**, 17–22.

37. Gline,l K., Thebault, P., Humblo, V., Pradier, C. M., and Jouenne, T. (2012). Antibacterial surfaces developed from bio-inspired approaches, *Acta Biomater,* **8**, 1670–1684.

38. Gong, P., Li, H., He, X., Wang, K., Hu, J., Zhang, S., and Yang, X. (2007). Preparation and antibacterial activity of Fe_3O_4@Ag nanoparticles, *Nanotechnology*, **18**, 604–611.

39. Gould, C. V., Umscheid, C. A., Agarwal, R. K., Kuntz, G., and Pegues, D. A. (2010). Healthcare infection control practices advisory committee. Guideline for prevention of catheter-associated urinary tract infections 2009, *Infect. Control Hosp. Epidemiol.*, **31**, 319–326.

40. Gutiérrez, D., Martínez, B., Rodríguez, A., and García, P. (2010). Isolation and characterization of bacteriophages infecting *Staphylococcus epidermidis, Curr. Microbiol.*, **61**, 601–608.

41. Gutiérrez, D., Martínez, B., Rodríguez, A., and García, P. (2012). Genomic characterization of two *Staphylococcus epidermidis* bacteriophages with anti-biofilm potential, *BMC Genomics*, **13**, 228.

42. Hall-Stoodley, L., Costerton, J. W., and Stoodley, P. (2004). Bacterial biofilms: From the natural environment to infectious diseases, *Nat. Rev. Microbiol.*, **2**, 95–108.

43. Hasan, J., Crawford, R. J., and Ivanova E. P. (2013), Antibacterial surfaces: The quest for a new generation of biomaterials, *Trends Biotechnol.*, **31**, 295–304.

44. Heijkants, R. G. J. C. (2006). Nanotechnology delivers microcoatings, *Med. Device Technol.*, **17**, 14–16.

45. Helmus, M. N., Gibbons, D. F., and Cebon, D. (2008). Biocompatibility: meeting a key functional requirement of next-generation medical devices, *Toxicol. Pathol.*, **36**, 70–80.

46. Hetrick, E. M., and Schoenfisch, M. H. (2006). Reducing implant-related infections: Active release strategies, *Chem. Soc. Rev.*, **5**, 780–789.

47. Hsu, S. H., Tseng. H. J., and Lin, Y. C. (2010). The biocompatibility and antibacterial properties of waterborne polyur-ethane-silver nanocomposites. *Biomaterials,* **31**, 6796–6808.

48. Ivanova, E. P., Hasan, J., Truong, V. K., Wang, J. Y., Raveggi, M., and Fluke, C., and Crawford, R. J. (2011). The influence of nanoscopically thin silver films on bacterial viability and attachment, *Appl. Microbiol. Biotechnol.*, **91**, 1149–1157.

49. Jaeblon, T. (2010). Polymethylmethacrylate: Properties and contemporary uses in orthopaedics, *J. Am. Acad. Otrhop. Surg.*, **18**, 297–305.

50. Kaali, P., Strömberg, E., and Karlsson S. (2011). Prevention of Biofilm Associated Infections and Degradation of Polymeric Materials Used in Biomedical Applications, in *Biomedical Engineering, Trends in Materials Scienc*e (Anthony Laskovski, ed.), ISBN: 978-953- 307-513-6.

51. Katsikogianni, M., and Missirlis, Y. F. (2004). Concise review of mechanisms of bacterial adhesion to biomaterials and of techniques used in estimating bacteria-material interactions, *Eur. Cell Mater.*, **8**, 37–57.

52. Kaur, H., Kumar, P., Ray, P., Kaur, J., and Chakraborti A. (2009). Biofilm formation in clinical isolates of group B streptococci from north India, *Microb. Pathog.*, **46**, 321–327.

53. Knetsch, M. L. W., and Koole, L. H. (2011). New strategies in the development of antimicrobial coatings: the example of increasing usage of silver and silver nanoparticles. *Polymers,* **3**, 340–366.

54. Kokare, C. R., Chakraborty, S., Khopade, A. N., and Mahadik, K. R. (2009). Biofilm: Importance and applications, *Ind. J. Biotechnol.*, **8**, 159–168.

55. Kutateladze, M., and, Adamia, R. (2010). Bacteriophages as potential new therapeutics to replace or supplement antibiotics, *Trends Biotechnol.*, **28**, 591–595.

56. Lellouche, J., Friedman, A., Gedanken, A., and Banin, E. (2012). Antibacterial and antibiofilm properties of yttrium fluoride nanoparticles, *Int. J. Nanomedicine,* **7**, 5611–5624.

57. Lönn-Stensrud, J., Landin, M. A., Benneche, T., Petersen, F. C., and Scheie, A. A. (2009). Furanones, potential agents for preventing *Staphylococcus epidermidis* biofilm infections? *J. Antimicrob. Chemother.*, **63**, 309–316.

58. Manefield, M., Rasmussen, T. B., Henzter, M., Andersen, J. B., Steinberg, P., Kjelleberg. S., and Givskov, M. (2002). Halogenated furanones inhibit quorum sensing through accelerated LuxR turnover, *Microbiology,* **148**, 1119–1127.

59. McLean, R. J., Hussain, A. A., Sayer. M., Vincent, P. J., Hughes, D. J., and Smith, T. J. N. (1993). Antibacterial activity of multilayer silver-copper surface films on catheter material, *Can. J. Microbiol.*, **39**, 895–899.

60. Mitick-Dineva, N., Wang, J., Truong, V. K., Stoddart, P., Malherbe, F., Crawford, R. J., and Ivanova E. P. (2009). *Escherichia coli*, Pseudomonas aeruginosa, and *Staphylococcus aureus* attachment patterns on glass surfaces with nanoscale roughness, *Curr. Microbiol.*, **58**, 268–273.

61. Murata, H., Koepsel, R. R., Matyjaszewski, K., and Russell A. J. (2007). Permanent, non-leaching antibacterial surfaces–2: how high density cationic surfaces kill bacterial cells, *Biomaterials,* **28**, 4870–4879.

62. Nomura, S., Lundberg, F., Stollenwerk, M., Nakamura, K., and Ljungh, A. (1997). Adhesion of staphylococci to polymers with and without immobilized heparin in cerebrospinal fluid, *J. Biomed. Mater. Res.*, **38**, 35–42.

63. Norowski, P. A., and Bumgardner, J. D. (2009). Biomaterial and antibiotic strategies for peri-implants, *J. Biomed. Mater. Res. B Appl. Biomater.*, **88B**, 530–543.

64. Nowatzki, P. J., Koepsel, R. R., Stoodley, P., Min, K., Harper, A., Murata, H., Donfack, J., Hortelano, E. R., Ehrlich, G. D., and Russell. A. J. (2012). Salicylic acid-releasing polyurethane acrylate polymers as anti-biofilm urological catheter coatings, *Acta Biomater.,* **8**, 1869–1880.

65. O'Grady, N. P., Alexander, M., Burns, L. A., Dellinger, E. P., Garland, J., Heard, S. O., Lipsett, P. A., Masur, H., Mermel, L. A., Pearson, M. L., Raad, I. I., Randolph, A. G., Rupp, M. E., and Saint, S. (2011). Guidelines for the prevention of intravascular catheter-related infections, *Clin. Infect. Dis.,* **52**, 162–193.

66. Otto, M. (2009). Staphylococcus epidermidis, the "accidental" pathogen, *Nat. Rev. Microbiol.,* **7**, 555–567.

67. Pavesio, A., Renier, D., Cassinelli, C., and Morra, M. (1997). Anti-adhesive surfaces through hyaluronan coatings. *Med. Device Technol.,* **8**(20–1), 24–27.

68. Rai, M., Yadav, A., and Gade, A. (2009). Silver nanoparticles as a new generation of antimicrobials, *Biotechnol. Adv.*, **27**, 76–83.

69. Ren, D., Bedzyk, L. A., Ye, R. W., Thomas, S. M., and Wood, T. K. (2004). Differential gene expression shows natural brominated furanones interfere with the autoinducer-2 bacterial signaling system of *Escherichia coli*, *Biotechnol. Bioeng.*, **8**8, 630–642.

70. Roe, D., Karandika, B., Bonn-Savage, N., Gibbin, B., and Roullet, J. B. (2008). Antimicrobial surface functionalization of plastic catheters by silver nanoparticles, *J. Antimicrob. Chemother.*, **61**, 869–876.

71. Roosjen, A., van der Mei, H. C., Busscher, H. J., and Norde, W. (2004). Microbial adhesion to poly(ethyle oxide) brushes: Influence of polymer chain length and temperature, *Langmuir,* **20**, 10949–10955.

72. Rosenthal, V. D., Maki, D. G., and Graves, N. (2008). The international nosocomical infection control consortium (INICC): Goals and objectives, description of surveillance methods, and operational activities, *Am. J. Infect. Control.*, **36**, 1–12.

73. Ryan, E. M., Gorman, S. P., Donnelly, R. F., and Gilmore, B. F. (2011). Recent advances in bacteriophage therapy: How delivery routes, formulation, concentration and timing influence the success of phage therapy, *J. Pharm. Pharmacol.*, **63**, 1253–1264.

74. Samuel, U., and Guggenbichle, J. P. (2004). Prevention of catheter-related infections: The potential of a new nano-silver impregnated catheter, *Int. J. Antimicrob. Agents*, **23**, 75–78.

75. Satishkumar, R., Sankar, S., Yurko, Y., Lincourt, A., Shipp, J., Heniford, B. T., and Vertegel, A. (2011). Evaluation of the antimicrobial activity of lysostaphin-coated hernia repair meshes, *Antimicrob. Agent Chemother.*, **55**, 4379–4385.

76. Scheuerman, T. R., Camper., A. K., and Hamilton, M. A. (1998). Effects of substratum topography on bacterial adhesion, *J. Colloid Interf. Sci.*, **208**, 23–33.

77. Schierholz, J. M., and Beuth J. (2001). Implant Infections: A haven for opportunistic bacteria, *J. Hosp. Infect.*, **49**, 87–93.

78. Schillaci, D., Arizza, V., Dayton, T., Camarda, L., and Di Stefano, D. (2008). In vitro anti-biofilm activity of Boswellia spp. oleogum resin essential oils, *Lett. Appl. Microbiol.*, **47**, 433–438.

79. Schreiber, F. (2000). Structure and growth of self-assembling monolayers, *Prog. Surf. Sci.*, **65**, 151–256.

80. Senaratne, W., Andruzzi, L., and Ober, C. K. (2005). Self-assembled monolayers and polymer brushes in biotechnology: Current applications and future perspectives, *Biomacromol*, **6**, 2427–2448.

81. Shah, N. J., Hong, J., Hyder, N., and Hammond, P. T. (2012) Osteophilic multilayer coatings for accelerated bone tissue growth, *Adv. Mater.*, **24**, 1445–1450

82. Shintani, H. (2004). Modification of medical device surface to attain anti-infection, *Trends Biomater. Artif. Organs,* **18**, 1–8.

83. Sioshansi, P. (1994). New processes for surface treatment of catheters, *Artif. Organs*, **18**, 266–271.

84. Sousa, C., Henriques, M., and Oliveira, R. (2011). Mini-review: Antimicrobial central venous catheters: Recent advances and strategies, *Biofouling*, **27**, 609–620.

85. Sousa, C., Henriques, M., Teixeira, P., and Oliveira, R. (2009). Reduction of *Staphylococcus epidermidis* adhesion to indwelling medical devices: a simple procedure, *Br. J. Biomed. Sci.*, **65**, 184–190.

86. Sousa, C., Teixeira, P., Bordeira, S., Fonseca, J., and Oliveira, R. (2008). *Staphylococcus epidermidis* adhesion to modified polycarbonate surfaces: Gold and SAMs coated, *J. Adhes. Sci. Technol.*, **22**, 675–686.

87. Sulakvelidze, A., and, Kutter, E. (2005). Bacteriophage therapy in humans. In *Bacteriophages: Biology and Application*. (Kutter, E., Sulakvelidze, A., ed.), pp. 381–436. Boca Raton, CRC Press.

88. Takenaka, S., Tonoki, T., Taira, K., Murakami, S., and Aoki, K. (2007). Adaptation of Pseudomonas sp. strain 7-6 to quaternary ammonium compounds and their degradation via dual pathways, *Appl. Environ. Microbiol.*, **73**, 1797–1802.

89. Tegoulia, V. A., and Cooper, S. L. (2002). *Staphylococcus aureus* adhesion to self-assembled monolayers: Effect of surface chemistry and fibrinogen presence, *Colloids. Surf. B Biointerf.*, **24**, 217–228.

90. Tenke, P., Riedl, C. R., Jones, G. L., Williams, G. J., Stickler, D., and Nagy, E. (2004). Bacterial biofilm formation on urologic devices and heparin coating as preventive strategy, *Int. J. Antimicrob. Agents*, **23**, 67–74.

91. Tiller, J. C., Liao, C. H., Lewis, K., and Klibanov, A. M. (2001). Designing surfaces that kill bacteria on contact, *Proc. Natl. Acad. Sci. U. S. A.*, **98**, 5981–5985.

92. Tran, P. A., and Webster, T. J. (2013). Antimicrobial selenium nanoparticle coatings on polymeric medical devices,. *Nanotechnology*, **24**, 155101.

93. Tsai, P. W., Yang, C. Y., Chang, H. T., and Lan, C. Y. (2011). Human antimicrobial peptide LL-37 inhibits adhesion of Candida albicans by interacting with yeast cell-wall carbohydrates, *PLoS ONE*, **6**, e17755.

94. Ulman, A. (1996). Formation and structure of self-assembled monolayers, *Chem. Rev.*, **96**, 1533–1554.

95. Wang, Q., and Webster, T. J. (2012). Nanostructured selenium for preventing biofilm formation on polycarbonate medical devices, *J. Biomed. Mater. Res.*, **100A**, 3205–3210.

96. Webb, J. C. J., and Spencer, R. F. (2007). The role of polymethylmethacrylate bone cement in modern orthopaedic surgery, *J. Bone Joint Surg.*, **89-B**, 851–857.

97. Willcox, M. D., Hume, E. B., Aliwarga, Y., Kumar, N., and Cole, N. (2008). A novel cationic-peptide coating for the prevention of microbial colonization on contact lenses, *J. Appl. Microbiol.*, **105**, 1817–1825.

98. Yue, I. C., Poff, J., Cortes, M. E., Sinisterra, R. D., Faris, C. B., Hildgen, P., Langer, R., and Shastri, V. P. (2004). A novel polymeric chlorhexidine delivery device for the treatment of periodontal disease, *Biomaterial*, **25**, 3743–50.

99. Zhao, L., Chu, P. K., Zhang, Y., and Wu, Z. (2009). Antibacterial coatings on titanium implants, *J. Biomed. Mater. Res. B Appl. Biomater.*, **91B**, 470–480.

100. Zodrow, K. R., Schiffman, J. D., and Elimelech, M. (2012). Biodegradable polymer (PLGA) coatings featuring cinnamaldehyde and carvacrol mitigate biofilm formation, *Langmuir*, **28**, 13993–13999.

Index

acid-buffering 252–253
acrylic acid 11, 13, 15, 32–33, 41, 239–241, 249, 253
active pharmaceutical ingredient (API) 25
adhesion
  anisotropic 85
  cell-to-cell 3–4, 8–9
  principles of 4–5, 7
  reversible 57, 138, 168
  shear 91–92, 141
  tissue 208–209, 211–212, 214–216, 220
adhesion energy 59
adhesion force 31–32, 57, 59, 92, 184, 195
adhesion prevention 273
adhesion strength 69, 71, 140, 170, 174, 181–183, 209, 211, 213–216, 220–221
adhesive coacervates 123
adhesive hydrogels 40, 69–70, 140
adhesive pads 86, 102–103
adhesive plaques 50, 52, 151–152
adhesive polymers 118, 132, 136, 142
adhesive protein 53, 56, 74, 110, 116–120, 122, 131
adhesive sealants 203, 215, 221
adhesives
  dental 15, 74
  directional 183, 185
  sensitive 14, 103
  wet 10, 14
AFM, *see* atomic force microscopy
albumin 15, 32, 40, 214–216, 250
algae 146, 168–169, 173, 176, 221
  brown 168–169
algal glue 168–169
algal glue mimetics 167–176
alginate 37, 40, 55–56, 137, 170, 172–174, 213, 218, 254
alginate gel 172–174
antibacterial agents 276–277
antibacterial surfaces 150, 274–275
antibiotics 218, 273, 277–278, 280, 282–284
antifouling polymers 72, 135, 144–145, 148–149
antimicrobial agents 271, 273, 278, 280, 284
antimicrobial peptides 271, 273, 277–278, 280, 283–284
API, *see* active pharmaceutical ingredient
atom transfer radical polymerization (ATRP) 72, 147
atomic force microscopy (AFM) 35–36, 42, 50, 54, 56–57, 60, 62–63, 74, 93, 115
ATRP, *see* atom transfer radical polymerization

β-keratin 88–89, 91, 97–98, 101–102, 182, 185
bacterial adhesion 271–278, 280, 284–285

prevention of 275, 280
bacteriophages 273, 281
bio-inspired adhesive polymers 118–119
bio-inspired surfaces 181–199
bioadhesion
characterization of 23–42
examples of 4, 8–15
principles of 1–21
bioadhesive films 27, 258–259
peel strength of 28, 259–260
bioadhesive ophthalmic drug 252
bioadhesive polymers 242, 249, 253, 255
bioadhesive retentive system 255
bioadhesive systems 26–27, 235–268
bioadhesive systems for drug delivery 235–260
bioadhesiveness 40, 42
bioadhesives 4, 14, 16, 49, 63, 203–222, 252
biofilm formation 269, 271–274, 277, 280–285
biofilms 247, 271, 279, 282–284
BioGlue 15, 206, 214–215, 220
bioinspired synthetic adhesives 138–139, 141, 143
biologic scaffolds, adhesive-coated 142
biological substrates 132–133, 135, 236, 242
biomimetic adhesives 71, 122, 140, 142
biomimetic algal glue 172–173, 175
biomimetic glue 173, 175–176
biomimetic materials 67–68, 74
bonds
bioadhesive 7, 25
strong adhesive 8, 11
bone 4, 16, 113, 123, 132, 149, 269
bovine thrombin 207–208, 210
byssal plaque 51–52, 54, 57, 64–65, 67
byssal threads 51, 70
byssus 49–50, 52, 55, 67
byssus threads 64, 134, 151–152

cadherins 9–10
catechol 55, 63, 68–74, 119, 122, 133–142, 145–147, 150–151
adhesive 138–140
cell adhesion 3–4, 8–10, 148, 274
cell adhesion specificity 10
cells
basal 237–238, 249
goblet 236–238, 251
chitin 216
chitosan 11, 32, 36, 38, 40, 70–71, 137, 206, 216–218, 240–241, 250–251
chitosan-based sealants 216–217
circular dichroism 56
CLSM, *see* confocal laser scanning microscopy
collagen 10, 15, 51–52, 54, 123, 205, 208, 210–213, 215–216, 218, 220
collagen sealants 209–212
colon 253, 255–256
colonization 269, 272–273, 275, 281, 285
confocal laser scanning microscopy (CLSM) 35, 113–114
corneal epithelium 238, 251

delivery systems
bioadhesive 249, 254, 256
mucoadhesive 29, 245–246, 248, 254–255
oral mucoadhesive 245–246
dentine 4, 15–16

devices, prosthetic 283–284
3,4-dihydroxyphenylalanine (DOPA) 52–57, 60–72, 110, 116–122, 131–134138, 145–147, 151, 211, 221
DOPA, *see* 3,4-dihydroxyphenylalanine
DOPA bidentate hydrogen bonding 61–62
DOPA chemistry 62, 68, 132
DOPA-deficient foot proteins 65
DOPA ligands 55, 65–66
DOPA-mediated underwater adhesion of mussel proteins 61
DOPA oxidation 60–61
DOPA residues 57, 116, 120, 135, 211
DOPA side chain 131–133, 135, 151
dopamine 51, 68, 71–72, 131–133, 135, 137, 141, 146–147, 149–151
dopaquinone 54, 60, 63–64, 66
drug delivery
  buccal 246
  mucoadhesive 14, 24–25, 237, 239, 260
drug delivery systems
  bioadhesive 26, 37, 235–236
  mucoadhesive 25, 237, 239, 260

enzymes 69, 69, 135, 151, 211, 236, 249, , 278–279, 282, 285
epithelium 237–238, 245, 248–249, 251–252, 255
essential oils 283
ethylene glycol 15, 32, 63, 73, 122, 137, 147, 213, 240, 244

fibrin-based adhesives 138, 205
fibrin-based sealants 206–209, 220
fibrin glue 69, 138–140, 142, 213, 218–219
fibrin sealants 14, 206–210, 219
  patient-derived 219
fucus glue 170, 172–173
Fucus serratus 170–172

gastric retention formulations (GRFs) 256
gastrointestinal tract 8, 10, 17, 24–25, 236, 255–256
gecko adhesion 85–88, 90, 102–103
gecko movement 88, 99–100
gecko pad 87–91, 94, 97, 99, 101–102
gecko setae 87, 182
gecko's foot 86–87, 94, 103
gelatine 14, 206, 209–214, 220
gelatine-based sealants 211
glutaraldehyde 206, 213–214
glycine 115–117, 120, 146, 151
GRFs, *see* gastric retention formulations

haemostasis 208–213, 217, 219–221
hybrid gecko- and mussel-inspired adhesive 143
hydrogels 15, 40, 69–71, 13, 140–141, 221, 244, 257
  mucoadhesive 257
hydrophilic surfaces 102, 173, 272

infections, medical device 273, 278, 285

interactions
  catechol–surface 133
  mucus–polymer 32
interfacial binding 131–133, 135, 140, 146

Kendall model 96, 98

laser tissue welding (LTW) 215
lectins 9–10, 16–17, 242, 260
LTW, *see* laser tissue welding

MAPs, *see* mussel adhesive proteins
marine mussels 68, 70, 73
metal oxide surfaces 133–134
microbial adhesion 269, 273, 285
microorganisms 17, 271–272, 275–276, 279–280, 282, 284–285
  adhesion of 17, 271
mucin 11, 31–34, 36–40, 63, 132, 236, 244, 246, 257–258
  viscosities of 31–32
mucin glycoproteins 236
mucoadhesion 4, 10–13, 24–25, 27–31, 33–34, 36–37, 40–42, 236–237, 242–245, 251, 253, 256–258
  factors affecting 236, 242–243, 245
  measurement of 30, 31, 38
mucoadhesion analysis 33
mucoadhesive polymers 11, 27, 32–33, 38, 40–41, 239, 241, 250–252, 260
mucoadhesiveness 26–27, 35, 40
mucosa 4, 8, 12, 30, 37–39, 42, 219, 236–239, 245–246, 248–249, 256
mucus 4, 10–13, 29–33, 36, 38–40, 42, 236–238, 240, 242, 245–246, 249–252, 255, 257, 274
  cervical 238, 252
mucus gel 12–13, 30, 32, 34, 236
mussel adhesion 50, 52, 54, 57, 60, 64
mussel adhesive proteins (MAPs) 74, 131–132
mussel adhesives 49–52
mussel byssus 49–50, 52, 55, 134
mussel foot 51–52, 54, 55, 59–60, 62–64, 66
mussel foot protein interaction 59, 64
mussel foot proteins 54–55, 60, 62–63
mussel-inspired antifouling coatings 144–145
Mytilus edulis 131–132, 147, 211

nasal mucosa 38, 219, 236, 239, 248–249

OEG, *see* oligomeric ethylene glycol
oligomeric ethylene glycol (OEG) 147
oral cancer, chemoprevention of 247–248
oral cavity 4, 37, 239, 245–246

peel-zone model 95–96, 98, 103
PEG-catechol 138–140
phloroglucinol 170, 173–175
Phragmatopoma californica 110, 116, 123
piezoelectric tubes 56–58
plaque proteins 51–53, 66

plaque–substratum interfaces 54, 64
PLL, *see* poly L-lysine
poly L-lysine (PLL) 65
polyacrylic acids 33, 41, 253
polybrene 32
polycarbophil 11, 250–251
polychaetes 109–110
polydopamine 135, 147–151
polydopamine-coated surfaces 149
polymer gels 258
polymer ligands 73
polymer mucoadhesion 36
polymeric backbone 241–242
polymerization 15–16, 56, 59, 69, 71–72, 122–123, 133, 135, 137, 145–147, 149
  surface-initiated 59, 145–147
polymers
  cross-linked 243
  linear 136
  mucoadhesive cationic 36
  mucoadhesive hydrophilic 25
  mussel-inspired DOPA-rich 63, 71
  phenolic 169–170
  thiolated 42, 241–242
polyphenols 169–170, 173, 175
polysaccharides 71, 117, 147–149, 168–169, 276
  sulphated 110, 115, 117
polystyrene 55, 63, 71, 135, 279
protein adsorption 55, 59, 120, 273, 275
proteins, polybasic 117–118
protein–substratum interactions 60

QACs, *see* quaternary ammonium compounds
quaternary ammonium compounds (QACs) 276
quinone 68–69, 72, 133, 135

recombinant proteins 119–120
RIP, *see* RNAIII-inhibiting peptide
RNAIII-inhibiting peptide (RIP) 280

Sabellaria alveolata 110–112, 114, 116
SAMs, *see* self-assembled monolayers
sealants
  bioadhesive 221
  collagen-based 209–210
selectins 9–10, 17
self-assembled monolayers (SAMs) 276
self-healing 68, 70–71, 74, 134, 152
SFA, *see* surface forces apparatus
silica 58, 62, 120, 122
silver nanoparticles 73, 150
SIP, *see* surface-initiated polymerization
SLS, *see* static light scattering
small intestine 4, 30, 236, 255
SPR, *see* surface plasmon resonance
Staphylococcus aureus 217, 251
static light scattering (SLS) 170
surface forces apparatus (SFA) 50, 57–60,
surface-initiated polymerization (SIP) 59, 145–147
surface plasmon resonance (SPR) 37–38, 55, 170, 257
surgical adhesives 204, 220
surgical sealants 203–206, 209, 213, 219–221
sutures 14, 138, 140, 143, 203–204, 220

synthetic adhesive polymers 136–137
  architectures of 136–137
synthetic adhesives 69, 132, 151, 167, 169, 205
synthetic polymers 122, 132, 136–137, 270
tissue sealants 4, 14, 214
tissues, mucosal 23–24, 27, 34, 40, 134, 235
titanium 73, 120–122, 134–135, 147, 276
Tokay Gecko 87–89, 91, 182
toluidine blue O (TBO) 245

tablets
  bioadhesive vaginal 253
  mucoadhesive buccal 246
TBO, *see* toluidine blue O
thrombin 14–15, 206–212, 220
tissue adhesion strength 211, 214–216, 220
tissue adhesives 3, 15, 138, 167, 176

underwater adhesion 50, 52, 61, 63, 110, 122, 131

wet adhesion 50, 57, 63, 68–69, 74, 86, 144
wound dressings 3–4, 14